NONLINEAR MIXTURE MODELS

A Bayesian Approach

NONLINEAR MIXTURE MODELS

A Bayesian Approach

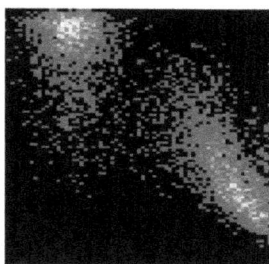

Tatiana Tatarinova

University of Southern California, USA

Alan Schumitzky

University of Southern California, USA

Imperial College Press

Published by

Imperial College Press
57 Shelton Street
Covent Garden
London WC2H 9HE

Distributed by

World Scientific Publishing Co. Pte. Ltd.
5 Toh Tuck Link, Singapore 596224
USA office: 27 Warren Street, Suite 401-402, Hackensack, NJ 07601
UK office: 57 Shelton Street, Covent Garden, London WC2H 9HE

Library of Congress Cataloging-in-Publication Data
Tatarinova, Tatiana V.
   Nonlinear mixture models : a bayesian approach / Tatiana Tatarinova, University of Glamorgan,
UK, Alan Schumitzky, University of Southern California, USA.
      pages cm
   Includes bibliographical references and index.
   ISBN 978-1-84816-756-8 (hardcover : alk. paper)
   1. Markov processes. 2. Bayesian statistical decision theory. 3. Nonparametric statistics.
4. Multivariate analysis. I. Schumitzky, Alan. II. Title.
   QA274.7.T38 2015
   519.2'33--dc23
                              2014038898

British Library Cataloguing-in-Publication Data
A catalogue record for this book is available from the British Library.

To our parents: Natalie and Valerii Tatarinov and Dorothy and Abe Schumitzky

Contents

List of Tables

List of Figures

Acknowledgments

We would like to thank personally the many friends, colleagues and students who have helped to write this book through great discussion and ideas: Roger Jelliffe, Michael Neely, Jay Bartroff, David Bayard, Michael Van Guilder, Walter Yamada, David D'Argenio, Xiaoning Wang, Carolyn Lindberg, Michael Christie, Anna-Michelle Christie, Alyona Kryschenko, Hannah Garbett and Farzana Rahmann. We would also like to thank those statisticians and mathematicians whose works inspired many of the ideas in this book: Christian Robert, the late George Casella, Mathew Stephens, Jon Wakefield, Chuan Zhou, Sujit Sahu, Russell Cheng, Allain Mallet, Peter Müller, Gary Rosner, Hedibert Freitas Lopes, Sylvia Fruwirth-Schnatter and Adrian Smith.

Special thanks go to Alyona Kryschenko and Carolyn Lindberg who helped proofread the book; and to Catharina Weijman, our extremely helpful editor from Imperial College Press. Of course there are others names we should have mentioned. Please accept our regrets.

Because the authors were practically never in the same place at the same time, this book took much longer to finish than was expected. And we are greatly indebted to our spouses, Michael and Carolyn, for bearing with us and for their patience and understanding.

Finally, this book has been supported in part by NIH grants GM068968, EB005803 and EB001978, NIH-NICHD grant HD070996, Royal Society grant TG103083, and by the University of Southern California and University of South Wales.

Chapter 1

Introduction

In this book we study Bayesian analysis of nonlinear hierarchical mixture models with a finite but possibly unknown number of components. Our approach will be based on Markov chain Monte Carlo (MCMC) methods. Application of our methods will be directed to problems in gene expression analysis and population pharmacokinetics in which these nonlinear mixture models arise very naturally. For gene expression data, one application will be to determine which genes should be associated with the same component of the mixture (clustering problem). For population pharmacokinetics, the nonlinear mixture model, based on previous clinical data, becomes the prior distribution for individual therapy. Bayesian analysis of prediction and control can then be performed. From a mathematical and statistical point of view, these are the problems analyzed in this book:

(1) Theoretical and practical convergence problems of the MCMC method.
(2) Clustering problems.
(3) Determination of the number of components in the mixture.
(4) Computational problems associated with likelihood calculations.
(5) Comparison of parametric and nonparametric Bayesian approaches.

In the statistical literature, these problems have been mainly addressed in the linear case. However, the methods proposed to solve these problems in the linear case do not readily transfer to the nonlinear case. Developing methods for the nonlinear case is one of the main contributions of this book. We will now briefly discuss some of the more important aspects of these problems.

One of the mixture models we consider is described by the relations:

$$y_{ij} \sim N(\cdot|h_{ij}(\theta_i), \sigma^2),$$

$$\theta_i \sim \sum_{k=1}^{K} w_k N(\cdot|\mu_k, \Sigma_k),$$

(1.1)

for $i = 1, ..., N$; $j = 1, ..., n_i$, where y_{ij} denotes the jth measurement on the ith subject; $\{h_{ij}\}$ are known functions; σ^2 is a (possibly) unknown variance; $\{\theta_i\}$ are unknown subject-specific parameter vectors; $N(\cdot|\mu, \Sigma)$ denotes the multivariate normal distribution with mean vector μ and covariance matrix Σ; and $\sim$ means "distributed as". We will say that the model defined by Eq. (1.1) is *linear* if all of the functions $h_{ij}(\theta)$ are affine in θ. Otherwise, the model is *nonlinear*.

For our Bayesian approach, all the mixture model parameters

$$\Phi = (\{w_k, \mu_k, \Sigma_k; k = 1, ..., K\}, \sigma)$$

(1.2)

will be assigned appropriate prior densities.

In the population pharmacokinetic application, the word "subject" means individual and θ_i is a vector of subject-specific pharmacokinetic parameters such as clearances, rates of elimination, and volumes of distribution. In the gene expression data application, "subject" means gene and θ_i is a vector of subject-specific gene expression model parameters.

Our goal is to calculate the posterior density $\pi(\Phi|Y)$, where Φ is the unknown parameter vector defined by Eq. (1.2), and $Y = \{y_{ij}, i = 1, ..., N; j = 1, ..., n_i\}$ represents the data. For a known number of components K, the method we propose will be a combination of Gibbs and Metropolis sampling. We will prove convergence of the resulting Markov chain. In particular, we will prove that this method generates an irreducible Markov chain. This theoretically implies that the chain visits all parts of the posterior distribution surface. Nevertheless, in practice, the resulting Markov chain can get computationally trapped in a local region of the posterior surface and cannot escape. Actually, the probability of escape is so small that it is below the round-off error on the computer. Developing methods to avoid this problem will be proposed and evaluated.

The problem of determining which mixture component an observation is most likely to come from (clustering problem) is very important in analyzing gene expression data and will be studied in this book. It will be shown that in order to fix the problem of "trapping" the computed posterior distribution will be forced to be symmetric relative to all components of the mixture. Consequently, an observation will be equally likely to come from any component. Clustering must then be done outside of the MCMC method. Both traditional and new approaches are discussed.

The determination of the number of components in a mixture is of both theoretical and practical importance. In the classical approach, the mixture model is analyzed for a fixed number of components and a numerical measure such as deviance information criterion (DIC), Akaike information criterion (AIC) or Bayesian information criterion (BIC), is calculated. Another measure that is important in this book is based on the Kullback–Leibler (KL) distance. The number of components is varied and the model with the best numerical measure is chosen. In 1997, Richardson and Green developed a different approach where the number of components K is considered as a random parameter of the model; the resulting MCMC method then calculates the posterior distribution $\pi(\phi, K|Y)$. In 2000, Stephens proposed an important variation of the Richardson–Green method; this method also will be discussed in the book.

It turns out that these "trans-dimensional" methods have another important feature. By switching from one dimension to another, the chain has better mixing properties and the "trapping" problem is in some cases avoided. However, the methods of Richardson–Green and Stephens both require calculation of the model likelihood. For the nonlinear mixture models considered in this book, the likelihood calculation would require evaluation of high-dimensional integrals. This calculation is embedded in the Markov chain and would therefore need to be performed thousands of times. Consequently, an alternative approach was sought out. The KL distance measure we consider does not require numerical integration and can be evaluated within the Gibbs–Metropolis chain. (It is interesting and important to note that our Gibbs–Metropolis method itself does not require numerical integration.) A trans-dimensional method is then developed similar to Stephens' but replacing likelihood calculations with the weighted KL distance.

1.1 Bayesian Approach

The concept of the Bayesian approach was first introduced over 250 years ago by the Reverend Thomas Bayes (1701–1761), who proved a special case of what is now called Bayes' theorem. That his name lives on in modern statistics is a consequence of the publication of the paper "An essay towards solving a problem in the doctrine of chances", communicated after Bayes' death by Richard Price in 1763 [Bayes and Price (1763)].

A decade later, Pierre-Simon Laplace "re-invented" and generalized the Bayesian approach, possibly unaware of the previous works of Bayes. However, in the 19th century, the Bayesian approach essentially fell into disuse, at least among theoreticians.

The terms "frequentist" and "Bayesian" statistics were not used until 1937 [Neyman (1937); Fienberg (2006)]. Nevertheless, Bayesian thinking was maintained by a handful of prominent followers such as Bruno de Finetti and Harold Jeffreys. Works of Alan Turing, Leonard Savage, Dennis Lindley and others started the new wave of Bayesian thinking in the second half of the 20th century. First, the Bayesian approach was used by cryptographers and applied statisticians, while general academics considered flexible Bayesian rules [Chen (2006)] to be non-rigorous. Anti-Bayesian resistance subsided somewhat in the late 1980s to early 1990s, when efficient Bayesian computational methods were introduced (such as MCMC), along with the wider availability of more powerful computers. A short history of MCMC can be found in Robert and Casella (2011).

This led to an explosion of interest in Bayesian statistics which in turn led to more research on the topic. Additional applications were found and now are used in diverse areas such as astrophysics [Loreda (1995)], weather forecasting [Cofino *et al.* (2002)] and health care [Spiegelhalter *et al.* (2002)].

Bayes' theorem itself can be stated as follows: The probability distribution over the model parameters, called the "prior distribution" is usually denoted as $p(\phi)$. When data Y are collected, the information obtained from the new data can then be expressed as a "likelihood", which is proportional to the distribution of the observed data conditional on model parameters $p(Y|\phi)$. The prior distribution and likelihood are combined into a new distribution called the "posterior distribution" and Bayes' theorem is then:

$$p(\phi|Y) = \frac{p(Y, \phi)}{p(Y)} = \frac{p(Y|\phi)p(\phi)}{p(Y)} = \frac{p(Y|\phi)p(\phi)}{\int p(Y|\phi)p(\phi)d\phi}. \qquad (1.3)$$

1.2 Review of Applications of Mixture Models in Population Pharmacokinetics

Population pharmacokinetics is the study of how a drug behaves relative to an appropriate population of subjects. The pharmacokinetic (PK) model describes the distribution of the drug in the body and the drug effect given various forms of drug delivery and dose scheduling. In terms of the posterior distribution $\pi(\phi|Y)$, the vector ϕ represents mixture parameters of various PK variables such as clearances, elimination rate constants and volumes of distribution (Eqs. (1.1), (1.2)); the data Y can represent measurements of drug and effect over a population of subjects. The population PK problem is to determine the distribution of ϕ given the population data Y.

Pharmacokinetic models are inherently nonlinear (the simplest being the exponential decay of a drug). Mixture models arise when variations of subpopulations react to the drug differently, such as fast and slow metabolizers – a known genetic characteristic. In general, these subpopulations are known or hypothesized to exist, but unfortunately cannot be determined in advance. Consequently, the requirements for a nonlinear mixture model are in place.

There is an enormous amount of literature on population PK for non-mixture models, using both Bayesian and maximum likelihood methods, see Davidian and Giltinan (1995). More recently, the Bayesian analysis software program PKBUGS has been developed for non-mixture models in population PK [Lunn *et al.* (2002)]. It is an interesting fact that some of the earliest applications of modern MCMC came from population PK [Wakefield *et al.* (1994); Best *et al.* (1995)].

One of the first studies of mixture models in population PK was that of Mallet (1986), in which a nonparametric maximum likelihood distribution was derived for $\pi(\Phi|Y)$. This nonparametric distribution could be considered as a finite mixture. A theoretical study of the geometry of mixture distributions was developed earlier in Lindsay (1983). Various implementations of these methods were discussed initially in Lindsay (1983), Mallet (1986) and Schumitzky (1991), and more recently in Pilla *et al.* (2006), see also Sections 6.1 and 6.2.

The first Bayesian analyses of mixture models in population PK appear to be those of Wakefield and Walker (1997, 1998); Mueller and Rosner (1997); and Rosner and Mueller (1997). In these references, the nonparametric Mallet approach was given a Bayesian framework by the introduction of the so-called Dirichlet process prior. This led to a mixture model with an infinite number of components. More recent relevant publications for mixture models are found in Mengersen *et al.* (2011) and Lu and Huang (2014).

Aside from the references mentioned above, there are relatively few applications of Bayesian analysis for nonlinear mixture models with a finite number of components. And the majority of them come from population PK, see also Huang *et al.* (2006).

In Lopes (2000) and Lopes *et al.* (2003), nonlinear hierarchical finite mixture models of exactly the type we consider are studied. Trapping problems are avoided by placing constraints on the parameters. Determination of the number of components is done by a variation of the Richardson–Green trans-dimensional approach. Numerical integration for likelihood calculations is done by "bridge sampling". In this book we suggest alternative approaches to these problems.

In Zhou (2004), Wakefield *et al.* (2003) and Zhou and Wakefield (2005), nonlinear hierarchical finite mixture models are proposed in the context of time series for gene expression data. However, the models in theory could just as easily have been pharmacokinetic.

The above mentioned papers in this section treat *general* nonlinear finite mixture models. However, there are a few papers which just treat special cases that should be mentioned. Pauler and Laird (2000) analyze a special mixture model for applications to drug compliance. Riley and Ludden (2005) analyze a special mixture model to distinguish between a one- and two-compartment model. Of interest in Riley and Ludden (2005) is that they compare their Bayesian results with a maximum likelihood approach using the little-known "mixture" module of the well-known NONMEM program [Beal and Sheiner (1995)].

1.3   Review of Applications of Mixture Models to Problems in Computational Biology

The field of computational biology is relatively new, very competitive and extremely challenging. Scientists from various disciplines apply their knowledge to unveil mysteries of cells and genes. Numerous methods from old and established disciplines such as physics, mathematics, chemistry and computer science have propagated into the area of computational biology.

Statisticians have long developed a powerful methodology of mixture models that is applied to biostatistics, medicine, pharmacokinetics and pharmacodynamics. We would like to discuss applications of mixture models to various problems in computational biology. Many researchers agree that mixture models can provide a straightforward, convenient, flexible and effective way to model large complex datasets. "Mixture models, which can be viewed as clustering techniques, have become one of the most widely used statistical methods for the analysis of heterogeneous data. Mixture models enable flexible fitting of data arising from diverse fields such as astronomy, bioinformatics, genetics, hyperspectral imaging, medical imaging and minefield detection" [Xing and Pilla (2004)].

We start with a description of the application of the mixture models to the discovery of the patterns in groups of proteins or nucleic acids. As it was elegantly described Bailey (1995), the task was to find "the islands of similarity afloat in seas of randomness, often referred to as *motifs* in biological literature". Bailey and Gribskov (1998) developed an algorithm called MEME (multiple expectation maximization for motif elicitation) that models the probability distribution of motifs "rather than discovering the original" motif itself. Sequences in the nucleotide or protein alphabet are treated as observations. A motif is modeled "by a sequence of discrete random variables whose parameters give the probabilities of each of the different letters, occurring in each of the different positions in an occurrence of the motif". Background positions of letters were modeled as discrete random variables. MEME utilized finite mixture models, with one component describing the null distribution and another component describing the motif. The output of the program was the set of parameters describing the "motif component of the mixture". MEME is now considered to be the gold standard among motif-finding programs and several research groups are trying to challenge its leadership. Among the more recent developments, we would like to mention a paper by Keles *et al.* (2003) describing a new likelihood-based method, COMODE (constrained motif detection), for identifying significant motifs in nucleotide sequences. They applied prior knowledge of the structural characteristics of protein–DNA interactions to the motif search. Keles *et al.* (2003) showed that COMODE performs better than unconstrained methods in cases where structural information is available.

In population genetics, mixture models have been used for linkage analysis of complex diseases, as described in the paper by Devlin *et al.* (2002). The assumption is that some fraction α of the considered pedigree have a "linkage between markers in the vicinity of a disease gene," while the other $(1 - \alpha)$ families have random occurrences of the illness. The clustering algorithm of Devlin *et al.* (2002) partitions the population into two groups using analysis of affected sibling pairs. Comparison of the results with the competing algorithms showed that mixture models have several advantageous features that differentiate them from other methods: mixture models approximate the process generating the data and they "make explicit what attributes of data are of interest". Mixture models are used in works of Alexander *et al.* (2009); Stephens and Balding (2009); Li *et al.* (2010); Elhaik *et al.* (2014b).

The number of publicly available microarray datasets for various organisms has been growing exponentially in the last 15 years. This amount of data has attracted more and more attention from the community of biostatisticians, bioinformaticians and computational biologists.

Pan (2002) suggested applying mixture models to the analysis of microarray data. He addressed the fundamental issue of determining the differentially expressed genes "across two kinds of tissue samples or samples obtained under two experimental conditions". Let X_{ik} be the observed expression of a gene i from the subject k. The expression level can be presented as a sum of mean expression of control plants a_i, treatment-related expression b_k, $x_k = 0$ for control tissues and $x_k = 1$ for treatment tissues and experimental error ϵ_{ik}:

$$X_{ik} = a_i + b_i x_k + \epsilon_{ik}, \tag{1.4}$$

where $k = 1, ..., K$ denote control tissues and $k = K + 1, ..., 2K$ treatment tissues.

Wei Pan suggested that determining whether a gene has differential expression is equivalent to testing the following hypothesis:

$$H_0 : b_i = 0 \text{ vs. } H_1 : b_i \neq 0 \tag{1.5}$$

for all n genes on the array. For every gene, $K(K-1)/2$ values of statistic $Z_i = log(X_{ik}) - log(X_{im})$ are computed, where k, m are randomly selected tissues from treatment and control plants. At the next step, he approximated a distribution of Z_i under both hypotheses as a mixture of normals:

$$f^{(K)} = \sum_{k=1}^{K} w_k f_k(\cdot | \theta_k, \sigma_k^2). \tag{1.6}$$

He then fitted a normal mixture and used the likelihood-ratio test to test the null hypothesis. Pan demonstrated that the mixture models approach gives more reliable results as compared to other approaches, namely, the Wilcoxon rank test, t-test and regression modeling. Pan's explanation for these more reliable results was based on intrinsic properties of mixture models: one can directly estimate the null distribution and only a few weak assumptions have to be made regarding the data structure. He used a parametric bootstrap technique to choose a cut point in declaring statistical significance for identified genes while controlling for the number of false positives.

Efron *et al.* (2001) suggested the empirical Bayes method (EMB), which treated genes as originating from two groups, differentially and non-differentially expressed: $f(X) = p_0 f_0(X) + (1 - p_0) f_1(X)$. The posterior probability of the differential expression was estimated as $P_1(X | f_0, f_1, p_0) = (1 - p_0) f_1(X) / f(X)$. The approach produces a posteriori probabilities of activity differences for each gene. Use of the upper boundary to estimate $\hat{p}_0 = min_X f(X) / f_0(X)$ created a bias in estimating $\hat{P}_1(X)$. Do *et al.* (2003, 2005) developed an extension of Efron's method for analysis of differential gene expression using Bayesian mixture models. They proposed a nonparametric Bayesian approach to compute posterior probabilities of differential expression and developed the MCMC posterior simulation to generate samples from posterior distributions. Rahnenfuhrer and Futschik (2003) applied EBM to identify differentially expressed genes.

McLachlan (1999) and McLachlan *et al.* (2001) developed and released software named EMMIX-GENE that used a mixture of t-distributions, normal distributions, and factor analyzers to model gene expression. Mixture model parameters were estimated using a maximum likelihood approach. Their ideas were generalized by Martella (2006) who tested a two-component mixture model using a parametric boostrap approach. Broet *et al.* (2004) applied the mixture model-based strategy to compare multiple classes of genes in microarray experiments; they used a modified F-statistic and a mixture of Gaussian distributions to compute Bayesian estimates of misallocation measures (false discovery rate and false non-discovery rate) conditional upon data.

The impressive work of Ghosh (2004) and Ghosh and Chinnaiyan (2002) resulted in a COMPMIX algorithm based on the hierarchical mixture models and a combination of methods of moments and Expectation-Maximization (EM) algorithms to estimate model parameters. Ghosh suggested considering two-component mixtures (for two tissue types) on top of the three-component mixtures (up-regulated genes, down-regulated genes and unchanged genes). The program was used to analyze popular colon cancer data published earlier by Alon *et al.* (1999).

Medvedovic *et al.* (2004) developed a clustering program called GIMM (Gaussian infinite mixture modeling). This program takes into account information about in-between replicates variability. Medvedovic *et al.* considered a linear mixture model with an infinite number of components and used a Gibbs sampler to estimate model parameters (allocation variables for clustering). To overcome the intrinsic "stickiness" of the Gibbs sampler, Medvedovic *et al.* (2004) introduced a heuristic modification to the

Gibbs sampler based on the "reverse annealing" principle. Medvedovic *et al.* (2004) compared several clustering methods and discovered that algorithms based on mixture models outperformed heuristic algorithms. Moreover, mixture models "capable of capturing gene- and experiment-specific variability" performed better than others.

A group of University of Southern California Medical School scientists [Siegmund *et al.* (2004)] conducted a comparison of cluster analysis methods using DNA methylation data. They compared Gaussian, Bernoulli and Bernoulli-lognormal mixture models with traditional clustering methods and concluded that "model-based approaches had lower misclassification rates than the classical heuristic hierarchical cluster analysis". Additionally, they observed that the Bernoulli-lognormal mixture model consistently produced a lower misclassification rate than other mixture-based methods.

Wakefield *et al.* (2003) developed a method for modeling gene expression time series using a mixture model approach. They developed a method for probabilistic clustering of genes based on their expression profile. Wakefield *et al.* used a multivariate Gaussian mixture model to describe gene expression data and used a birth-death (BD) MCMC method [Stephens (2000b)] to find the optimal number of mixture components. One crucial difference between the approach of Wakefield *et al.* and other methods is the latter did not acknowledge time-ordering of the data. More recently Tatarinova (2006); Tatarinova *et al.* (2008) and, independently, Li (2006) and De la Cruz-Mesía *et al.* (2008) developed model-based clustering methods that group individuals according to their temporal profiles of gene expression using nonlinear Bayesian mixture models and maximum likelihood approaches. These approaches reconstruct underlying time dependence for each gene and provide models for measurement errors and noise. The goal is to cluster parameters of nonlinear curves, not the observed values. Hence, these methods can efficiently handle missing data as well as irregularly spaced time points. Model-based clustering has some inherent advantages compared to non-probabilistic clustering techniques [Li (2006)]. Probabilistic model-based clustering methods provide estimations of the distribution of parameters for each cluster as well as posterior probabilities of cluster membership. In this book, we generalize these ideas for nonlinear mixture models.

1.3.1 *Problems with mixture models*

It was shown by a number of researchers working in the area of finite mixture models (including Diebolt and Robert (1994), Richardson and Green (1997), Stephens (1997a), Fruhwirth-Schnatter (2001), Cappé *et al.* (2003)) that the intrinsic unidentifiability of the models, originating from the symmetry of the posterior distribution, makes analysis difficult. Artificial measures such as introducing additional identifiability constraints create a bias toward the constraint. Development of the BDMCMC and reversible jump MCMC [Stephens (1997a); Cappé *et al.* (2003)], which attempted to take advantage of the properties of the posterior distributions instead of fighting them, was a major breakthrough that allowed us to analyze consistently many datasets. However, as pointed out by Fruhwirth-Schnatter (2001), "the Gibbs sampler does not explore the whole unconstrained parameter space, but tends to stick at the current labeling subspaces with occasional, unbalanced switches between the labeling subspaces", producing non-symmetric posterior distributions. This problem becomes especially pronounced for nonlinear mixture models. Fruhwirth-Schnatter proposed using a random permutation sampler (RPS) that permutes the component labels to achieve properly symmetric posterior distributions. In addition, since BDMCMC and reversible jump MCMC involve computation of the likelihood at every dimension-changing step, the likelihood can be difficult to numerically calculate for some complex models. Sahu and Cheng (2003) have developed a KL distance-based approach to the selection of the number of mixture components. This calculation does not involve numeric integration and it was extremely tempting to combine the strategies together in search of the perfect method of analysis of the finite mixture models.

These were our starting points and motivations for the development of the KLMCMC (Kullback–Leibler Markov chain Monte Carlo) algorithm. We have attempted to take the best from previously developed methods and combine them into a mathematically consistent approach.

1.4 Outline of the Book

In Chapter 2 *Mathematical Description of Nonlinear Mixture Models*, we first provide a brief introduction to the MCMC methods, define basic notions and describe some popular methods and algorithms. Next we discuss mixture models and concentrate on an important special case of nonlinear normal mixture models as defined in Eq. (1.1). If $h_i(\theta_i) \equiv \theta_i$ we get a linear mixture model. A popular example is the Eyes dataset, containing "peak sensitivity wavelengths for individual microspectrophotometric records on a small set of monkey eyes" [Bowmaker *et al.* (1985)]. This simple example is perfect for illustration of the major concepts and methods, and will be used throughout this book. The chapter is concluded with discussion of the Gibbs sampler, the WinBUGS software (Section 2.3) and selection of prior distributions (Section 2.4).

In Chapter 3 *Label Switching and Trapping*, we discuss issues and methods of analysis of convergence of a Markov chain. We start (Section 3.1) with description of the label-switching phenomenon and its relationship to convergence of the Markov chain. We describe several methods for convergence analysis: Raftery–Lewis, Gelman–Rubin, Brooks–Gelman and Geweke, autocorrelations, posterior symmetry. We then discuss their applicability in identifying problems with the existence of trapping states (Section 3.2).

In the following sections we present three existing approaches: random permutation sampler (Section 3.1), re-parametrization (introduction of identifiability constraints, Section 3.4), and relabeling (choosing a permutation of component labels that minimizes the posterior expectation of a loss function, Section 3.5). In Section 3.3.1, we discuss the possibility of using RPS as a post-processing step.

In Chapter 4 *Treatment of Mixture Models with an Unknown Number of Components*, we discuss methods and problems related to the treatment of mixture models with an unknown number of components. The scope of the problem and terminology are outlined in Section 4.1. In Section 4.2, we discuss the distance-based approach of Sahu and Cheng (2003) to find the smallest number of components that explain the structure of the data. This task is traditionally accomplished by the *reductive stepwise method* [Sahu and Cheng (2003)]. In the framework of this method, one starts with the $k = k_0$ component mixture, which adequately explains the observed data. Then k is iteratively reduced until the fit is no longer adequate. Reduction of the number of components is achieved by *collapsing* two of the k components. Adequacy of the fit is assessed using KL distance between probability densities f and g:

$$d(f, g) = \int f(x) log(\frac{f(x)}{g(x)}) dx. \qquad (1.7)$$

We apply this method to the normal mixture models, illustrated by the Eyes example and calculate a closed-form expression for a mixture of beta distributions.

In Section 4.3, we describe an alternate approach: BCMCMC algorithm developed by Stephens (2000b). Under the BDMCMC methodology, a number of components of the mixture change dynamically: new components are created (*birth*), or an existing one is deleted (*death*) and model parameters are then recomputed. We have applied BDMCMC to the analysis of the Eyes dataset and a more complex nonlinear normal mixture model. Due to the invariance of likelihood under relabeling of the mixture components, posterior distributions of model parameters are bimodal.

Section 4.4 is devoted to the description of an algorithm for the finite mixture analysis. The work of Sahu and Cheng (2003) has inspired us to merge the method of choosing the optimal number of components using the weighted KL distance, BDMCMC approach outlined by Stephens (2000b) and the random permutation sampler from Fruhwirth-Schnatter (2001). We apply the KLMCMC algorithm to the Eyes model and test the method for a wide range of parameters.

In Chapter 5 *Applications of BDMCMC, KLMCMC, and RPS*, we present several examples of how BDMCMC, KLMCMC, and RPS deal with the problem of the selection of the optimal number of mixture components. As examples we consider: the *Galaxy* dataset (Section 5.1), simulated nonlinear mixture model (Section 5.2), the two-component *Boys and Girls* dental measurements dataset (Section 5.3), a one-compartment PK model (Section 5.4), and a nonlinear gene expression model (Section 5.5).

In referring to the algorithms used we mean precisely: BDMCMC – Algorithm 4.1, KLMCMC – Algorithm 4.2, RPS – Algorithm 3.1, Stephens relabeling – Algorithm 3.2. Our goal in each case is to determine the number of mixture components and to estimate the unknown values of component-specific parameters.

In the previous chapters we studied parametric mixture models where the mixture components were continuous parameterized densities. In Chapter 6 *Nonparametric Methods*, we consider the nonparametric case where the mixture model is a sum of delta functions. We consider both maximum likelihood and Bayesian estimation.

The purpose of this chapter is to explain how the methods of nonparametric maximum likelihood (NPML) and nonparametric Bayesian (NPB) are used to estimate an unknown probability distribution from noisy data. Although this book is about the Bayesian approach, there are important connections between NPML estimation and NPB ideas.

In Section 6.1, we define the basic model used in this chapter. In Section 6.2, we show that the NPML estimation problem can be cast in the form of a primal-dual interior-point method. Then, details of an efficient computational algorithm are given. For NPML, nonlinear problems are no harder than linear problems. This is illustrated with the nonlinear PK model.

In Section 6.3, we define the Dirichlet process for NPB and discuss some of its most important properties. In Section 6.4, we describe the Gibbs sampler for the Dirichlet process and discuss some technical considerations. In Section 6.5, we give a number of NPB examples using the Gibbs sampler, including, among others, the common benchmark datasets: Galaxy, Thumbtacks, and Eye Tracking. In Section 6.6, various technical details about condensing the MCMC output are discussed: cleaning, clustering and dealing with multiplicity.

In Section 6.7, we describe the constructive definition of the Dirichlet process given by Sethuraman (1994), called the stick-breaking process. A natural truncation of the stick-breaking process reduces the Dirichlet process to a finite mixture model with a known number of components. Consequently, all the algorithms in Chapters 2–4 will apply to this case. In Section 6.8, examples are then given to compare results from the truncated stick-breaking process to the full Dirichlet process, including the important nonlinear PK model from earlier chapters. In all the above examples, WinBUGS or JAGS programs are provided. In Section 6.9, we show an important connection between NPML and NPB.

In Chapter 7 *Bayesian Clustering Methods*, we discuss applications of algorithms developed in the previous chapters to different datasets and present our KL clustering (KLC) method for treatment of data-rich problems. We start with a description of previously developed clustering methods and motivation for the development of this approach (Section 7.1). In Section 7.2, we describe the application of KLMCMC to gene expression time series. We describe the KLC algorithm in Section 7.3, and demonstrate utility of this method in Section 7.4. In Section 7.5, an application of the NPB method to prediction of transcription start sites is given. We conclude this chapter in Section 7.6.

Cover graphics were borrowed from a recent paper by Elhaik *et al.* (2014a). It shows a clear two-dimensional bimodal distribution of properties of coding regions in rice: fraction of cytosines and guanines in the third position of codons (GC_3) and gene-body methylation level.

Chapter 2

Mathematical Description of Nonlinear Mixture Models

In this chapter, we will discuss methods for calculating the posterior distribution $\pi(\Phi|Y)$ for a parameter vector Φ conditional on data Y, as well as functionals of this distribution $F(\pi) = \int f(\Phi)\pi(\Phi|Y)d\Phi$. We will concentrate on the extremely interesting and difficult case of nonlinear hierarchical mixture models. Until recently, there was no practical way to calculate $\pi(\Phi|Y)$ and $F(\pi)$ for such complex models. Development of the Markov chain Monte Carlo (MCMC) methodology has drastically changed this situation. In this chapter, we introduce fundamental notions of MCMC theory that we are going to use in this book (Section 2.1). In Section 2.2 we introduce general nonlinear normal hierarchical models. Section 2.3 is devoted to the description of Gibbs sampling in the framework of MCMC. This chapter is concluded with a discussion of selection of prior distributions for linear and nonlinear models (Section 2.4). We will see that the model defined by a Dirichlet process in Chapter 6 has much in common with the finite mixture model with a bounded but unknown number of components.

2.1 Fundamental Notions of Markov chain Monte Carlo

In this section we give informal definitions of the basic Markov chain concepts that will be needed later. For precise details, see for example Robert and Casella (2004).

Let $X_t, t = 0, 1, ..., n$ be a sequence of random vectors with values in a space E with both continuous and discrete components. X_t is a Markov chain if

$$P(X_{t+1} \in A|X_0, ..., X_t) = P(X_{t+1} \in A|X_t), \text{ for } A \subset E. \quad (2.1)$$

The Markov chain has an invariant (or stationary) distribution π if $X_t \sim \pi$

implies that $X_{t+1} \sim \pi$ for all t, where the symbol $\sim$ means "distributed as".

The Markov chain is *irreducible* if for all $A \subset E$ such that $\pi(A) > 0$ we have $P(X_t \in A$, for some $t > 0 | X_0 = x) > 0$ for all $x \in E$. What is practically important is the fact that an irreducible Markov chain at some point reaches any subset with positive measure.

The formulation of the ergodic theorem is taken from the book *Markov chain Monte Carlo in Practice* edited by Gilks, Richardson and Spiegelhalter (Chapter 3 in Gilks *et al.* (1996)).

Theorem 2.1. *Let* $\{X_t\}$ *be an irreducible Markov chain with invariant distribution* π *and let* f *be a real-valued function on* E, *such that* $E_\pi(|f|) < \infty$. *Then,*

$$P(lim_{n\to\infty}\frac{1}{n}\sum_{t=1}^{n}f(X_t) = E_\pi(f)|X_0 = x) = 1 \qquad (2.2)$$

for π-*almost all* $x \in E$.

The Markov chain is *aperiodic* if there does not exist a measurable partition $B = \{B_0, B_1, ..., B_{r-1}\}$ for some $r \geq 2$, such that $P(X_t \in B_{t(mod\{r\})} | X_0 = x \in B_0) = 1$, for all t. The next important theorem is taken from the same source (see Chapter 3 in Gilks *et al.* (1996)).

Theorem 2.2. *Let* $\{X_t\}$ *be an irreducible aperiodic Markov chain with invariant distribution* π. *Then,*

$$lim_{t\to\infty}P(X_t \in A|X_0 = x) = \pi(A) \qquad (2.3)$$

for π-*almost all* $x \in E$.

This theorem validates the procedure of letting the chain run for a sufficiently long time to get a realization from π.

In order to get n samples from the distribution π, we start with an arbitrary starting state X_0, and, after sufficient burn-in iterations, the Markov chain will "forget" about the starting state and the sample points $\{X_t\}$ will be generated from a distribution close to the stationary distribution of the Markov chain. Run the Markov chain for another n iterations, and, after discarding m burn-in iterations, the output of the Markov chain can be used to estimate $E(f(X))$ [Gilks *et al.* (1996)]:

$$E(f(X)) \approx \frac{1}{n}\sum_{t=m+1}^{m+n}f(X_t). \qquad (2.4)$$

There are several algorithms that construct the Markov chain with stationary distribution π. In 1953, inspired by earlier ideas of Enrico Fermi, physicist Nicholas Constantine Metropolis proposed a method for sampling an arbitrary probability distribution. This algorithm is currently known as the *Metropolis algorithm* and is regarded as one of the most important algorithms of the 20th century [Metropolis *et al.* (1953)].

In the Metropolis algorithm, at a "time" t, the next state X_{t+1} is chosen by sampling from a symmetric *proposal distribution* $q(Y|X) = q(X|Y)$. The candidate Y is accepted with probability

$$\alpha(X, Y) = min\left(1, \frac{\pi(Y)}{\pi(X)}\right). \tag{2.5}$$

Therefore,

$$\pi(X)P_{X,Y} = \pi(X)q(X|Y)\alpha(X,Y) = \pi(X)q(X|Y)min\left(1, \frac{\pi(Y)}{\pi(X)}\right)$$

$$= min\left(\pi(X)q(X|Y), \pi(Y)q(X|Y)\right) = min\left(\pi(X)q(X|Y), \pi(Y)q(Y|X)\right)$$

$$= \pi(Y)q(Y|X)\alpha(Y,X) = \pi(Y)P_{Y,X}, \tag{2.6}$$

where $P_{Y,X}$ and $P_{X,Y}$ are Markov chain transition probabilities. Since this expression is symmetric with respect to X and Y we conclude that the Markov chain for the Metropolis algorithm is "reversible". Algorithm 2.1 presents an outline of the Metropolis approach from Chapter 1 of Gilks *et al.* (1996).

Algorithm 2.1 Metropolis algorithm

Require: Initialize X_0, set $t = 0$.
Ensure: 1:
  Sample a point Y from $q(Y|X_t)$.
  Sample a point U from the uniform distribution $U[0, 1]$.
Ensure: 2:
  if $U \leq \alpha(X_t, Y)$
  set $X_{t+1} = Y$
  else set $X_{t+1} = X_t$
  Increment t.

An example of a symmetric proposal distribution is $q(Y|X) = q(|Y - X|)$, used in the *random walk* Metropolis algorithm. In 1970, Hastings generalized the Metropolis algorithm to work with non-symmetric proposal densities [Hastings (1970)]. He suggested using the following acceptance probability:

$$\alpha(X, Y) = min\left(1, \frac{\pi(Y)q(X|Y)}{\pi(X)q(Y|X)}\right). \qquad (2.7)$$

It is easy to see that the Markov chain for the Metropolis–Hastings algorithm is also reversible.

Metropolis developed another version of his algorithm [Metropolis *et al.* (1953)]. He suggested at each step of the algorithm to update just one component of X. Later, Geman and Geman (1984) suggested using a proposal distribution of the form $q^i(Y^i|X^i, X^{-i}) = \pi(Y^i|X^{-i})$, where $X^{-i} = \{X^1, ..., X^{i-1}, X^{i+1}, ..., X^r\}$ and $\pi(Y^i|X^{-i})$ is the full conditional distribution for i^{th} component of X conditioned on the most recent values of all other components of X. This "element-by-element" implementation of the Metropolis–Hastings algorithm is called *Gibbs sampling*. The acceptance probability is

$$\alpha(X^{-i}, X^i, Y^i) = min\left(1, \frac{\pi(Y^i|X^{-i})\pi(X^i|X^{-i})}{\pi(X^i|X^{-i})\pi(Y^i|X^{-i})}\right) \equiv 1, \qquad (2.8)$$

for all Y^i. In other words, candidates generated in the process of the Gibbs sampler are always accepted and "the Gibbs sampler consists purely in sampling from full conditional distributions" [Gilks *et al.* (1996)].

Now, let an element $X \in E$ be decomposed as follows: $X = (X^1, ..., X^r)$, where each element X^j can be either continuous or discrete. We would like to sample from $\pi(X)$, but this may be difficult. Assume that it is straightforward to sample from $\pi(X^i|X^{-i})$. The Gibbs sampler generates a Markov chain that has $\pi(X)$ as its invariance distribution as follows: take an arbitrary point $X_0 = (X_0^1, ..., X_0^r)$ and successively generate the random variables:

$$X_1^1 = \pi(X^1|X_0^2, X_0^3, ..., X_0^r)$$
$$X_1^2 = \pi(X^2|X_1^1, X_0^3, ..., X_0^r) \qquad (2.9)$$
$$\vdots$$
$$X_1^r = \pi(X^r|X_1^1, X_1^2, ..., X_1^{r-1})$$

These steps define a transition from X_0 to X_1. Iteration of this process generates a sequence $X_0, X_1, ..., X_t$ that is clearly a Markov chain by definition.

It is easy to show that the Gibbs sampler in Eq. (2.9) has invariant distribution $\pi(X)$. First, assume that $X_t = (X_t^1, ..., X_t^r) \sim \pi(X_t)$. From the definitions of the Gibbs sampler and conditional probability,

$$
\begin{aligned}
P(X_{t+1}^1, X_t^2, ..., X_t^r) &= P(X_{t+1}^1 | X_t^2, ..., X_t^r) P(X_t^2, ..., X_t^r) \\
&= \pi(X_{t+1}^1 | X_t^2, ..., X_t^r) \pi(X_t^2, ..., X_t^r) \qquad (2.10) \\
&= \pi(X_{t+1}^1, X_t^2, ..., X_t^r)
\end{aligned}
$$

Updating the second component results in

$$
\begin{aligned}
P(X_{t+1}^1, X_{t+1}^2, ..., X_t^r) &= P(X_{t+1}^2 | X_{t+1}^1, X_t^2, .., X_t^r) P(X_{t+1}^1, X_t^3, ..., X_t^r) \\
&= \pi(X_{t+1}^2 | X_{t+1}^1, X_t^3, ..., X_t^r) \pi(X_{t+1}^1, X_t^3, ..., X_t^r) \\
&= \pi(X_{t+1}^1, X_{t+1}^2, X_t^3, ..., X_t^r). \qquad (2.11)
\end{aligned}
$$

At the rth step,

$$
\begin{aligned}
P(X_{t+1}^1, X_{t+1}^2, ..., X_{t+1}^r) &= P(X_{t+1}^r | X_{t+1}^1, .., X_{t+1}^{r-1}) P(X_{t+1}^1, ..., X_{t+1}^{r-1}) \\
&= \pi(X_{t+1}^r | X_{t+1}^1, ..., X_{t+1}^{r-1}) \pi(X_{t+1}^1, ..., X_{t+1}^{r-1}) \\
&= \pi(X_{t+1}^1, X_{t+1}^2, ..., X_{t+1}^r). \qquad (2.12)
\end{aligned}
$$

If $X_t \sim \pi$, then $X_{t+1} \sim \pi$, i.e. π is by definition an invariant distribution for the Markov chain $\{X_t\}$.

2.2 Nonlinear Hierarchical Models

2.2.1 *One-component model*

Let us consider several individuals indexed by $i = 1, ..., N$. For each individual, we have n_i measurements indexed by $j = 1, ..., n_i$. Then y_{ij} denotes the jth measurement on individual i and let $Y_i = (y_{i1}, ..., y_{in_i})$. One of our basic structural models will be given by the nonlinear equation $Y_i \sim N(\cdot | h_i(\theta_i), \tau^{-1}\Omega_i)$, where $h_i(\theta_i) = (h_{i1}(\theta_i), ..., h_{in_i}(\theta_i))$ is a vector of nonlinear functions, θ_i is a subject-specific parameter vector, τ is a scale factor, Ω_i is a known subject-specific positive definite covariance matrix, and where, in general, $N(\cdot | \mu, \Sigma)$ is a multivariate normal distribution with mean vector μ and covariance matrix Σ, (A.2). All distributions denoted by (A.x) are defined in Appendix A.

The simple variance model assumed for Y_i could be considerably generalized. However, Eq. (2.13) is sufficient to describe all the examples considered in this book. We now add the probabilistic Bayesian framework.

First stage of the model:

$$Y_i \sim N(\cdot | h_i(\theta_i), \tau^{-1} \Omega_i). \tag{2.13}$$

We assume, conditional on θ_i and τ, that the $\{Y_i\}$ are independent. It follows that

$$\prod_{i=1}^{N} p(Y_i | \theta_i, \tau) = \prod_{i=1}^{N} N(Y_i | h_i(\theta_i), \tau^{-1} \Omega_i). \tag{2.14}$$

Conditional on μ and Σ, the θ_i are independent and identically distributed. Thus, the *second stage* of the model is

$$\theta_i \sim N(\cdot | \mu, \Sigma). \tag{2.15}$$

Assume τ, μ and Σ are mutually independent. Then the *third stage* of the model is

$$\begin{aligned} \tau &\sim G(\cdot | a, b), \\ \mu &\sim N(\cdot | \lambda, \Lambda), \\ \Sigma^{-1} &\sim W(\cdot | q, \Psi), \end{aligned} \tag{2.16}$$

where $G(\cdot | a, b)$ is a gamma distribution (A.3) and $W(\cdot | q, \Psi)$ is a Wishart distribution (A.7). The choice of hyper-parameters λ, q, Λ, a, b, and Ψ will be discussed in Section 2.4.

The model described by Eqs. (2.13)–(2.16) will be sufficient to handle our examples when the observations $\{Y_i\}$ are continuous random vectors from a normal distribution. But we will also later consider discrete cases.

In the first discrete case we will assume the density of Y_i is Poisson with mean parameter θ_i, i.e.

$$p(Y_i = k | \theta_i) = \frac{exp(-\theta_i) k^{\theta_i}}{k!}, \text{ for } k = 0, 1, \dots \tag{2.17}$$

In this case the natural (conjugate) distribution for θ_i is the gamma distribution $G(\cdot | a, b)$, where the hyper-parameters (a, b) may or may not have prior distributions.

In the second discrete case we will assume the density of Y_i is binomial based on m_i independent trials and where θ_i is the probability of success, i.e.

$$p(Y_i = k | \theta_i) = C_k^{m_i} \theta_i^k (1 - \theta_i)^{m_i - k} \text{ for } k = 0, 1, \dots, m_i. \tag{2.18}$$

In this case the natural (conjugate) distribution for θ_i is the beta distribution $Beta(\cdot | a, b)$, where the hyper-parameters (a, b) may or may not have prior distributions.

2.2.2 Mixture models

A sufficiently large class of interesting real-life problems can be effectively described using mixture models. Mixture models are generally used when observations are coming from one of K groups that have distinct properties and can be described by distinct probability distributions. In the framework of mixture models, one can estimate complex probability distributions that do not fit standard families of distributions. This technique is especially useful when measurements are pooled from fundamentally different sources, but those sources cannot be distinguished experimentally or visually.

A random variable θ is said to have a mixture distribution if the density $p(\theta)$ is of the form

$$p(\theta) = \sum_{k=1}^{K} w_k f_k(\theta), \qquad (2.19)$$

where the $\{f_k(\theta)\}$ are probability densities and where the $\{w_k\}$ are non-negative weights that sum to one. We will also write Eq. (2.19) as

$$\theta \sim \sum_{k=1}^{K} w_k f_k(\cdot) \qquad (2.20)$$

or

$$\theta \sim \sum_{k=1}^{K} w_k F_k(\cdot), \qquad (2.21)$$

where $F_k(\theta)$ is the probability distribution with density $f_k(\theta)$. In general, the notation $x \sim f$, or $x \sim F$, will mean that the random variable x has density f or distribution F.

More on notation: f will always be a probability density. F can be a probability distribution or a probability density as determined by context. Continuous densities will always be with respect to Lebesgue measure; discrete densities will always be with respect to the counting measure.

We will assume the component densities f_k and distributions F_k are of the form

$$f_k(\theta) = G(\theta|\varphi_k), \quad F_k(\theta) = G(\theta|\varphi_k),$$

where φ_k is a parameter vector lying in a subset of Euclidean space and $g(\cdot|\varphi_k)$ and $G(\cdot|\varphi_k)$ are probability density and distribution functions for each value of φ_k. In general, the weights $\{w_k\}$ and the parameters $\{\varphi_k\}$ will be unknown.

Now consider the case where the $\{\theta_i\}$ are a sequence of a random vectors that are independent and identically distributed (iid) with common distribution F so that

$$\theta \sim F(\cdot) = \sum_{k=1}^{K} w_k F_k(\cdot|\varphi_k).$$

In real-life problems, one is not usually able to observe the $\{\theta_i\}$ directly. Instead, we assume the observations $\{Y_i\}$ satisfy a relationship of the form

$$Y_i \sim F_Y(\cdot|\theta_i), \quad i = 1, ..., N, \tag{2.22}$$

where the $\{Y_i\}$ are vectors of independent but not necessarily identically distributed measurements with conditional distributions $\{F_Y(Y_i|\theta_i)\}$. To complete the Bayesian description of the model, we need to put a prior distribution F_w on the weights $\{w_k\}$ and a prior distribution F_φ on the parameters $\{\varphi_k\}$. So in summary the model is defined by the equations:

$$Y_i \sim F_Y(\cdot|\theta_i), \quad i = 1, ..., N,$$
$$\theta_i \sim \sum_{k=1}^{K} w_k F_\theta(\cdot|\varphi_k),$$
$$(w_1, ..., w_K) \sim F_w(\cdot), \tag{2.23}$$
$$(\varphi_1, ..., \varphi_K) \sim F_\varphi(\cdot).$$

In this book we will concentrate on four examples of the model of Eqs. (2.23). In the first example,

$$F_Y(Y_i|\theta_i) = N(Y_i - h_i(\theta_i), \tau_i^{-1}\Omega_i),$$
$$(\theta|\varphi_k) \sim N(\theta - \mu_k, \Sigma_k), \quad \varphi_k = (\mu_k, \Sigma_k),$$
$$(w_1, ..., w_K) \sim Dir(\cdot|\alpha), \quad \alpha = (\alpha_1, ..., \alpha_K),$$

where μ_k and Σ_k are independent, and $\mu_k \sim N(\cdot|\lambda_k, \Lambda_k)$, $\Sigma_k^{-1} \sim W(\cdot|q_k, \Psi_k)$; $N(\theta|\mu, \Sigma)$ is the multivariate normal distribution of θ with mean vector μ and covariance matrix Σ; $W(\cdot|q, \Psi)$ is the Wishart distribution (A.4); and $Dir(\cdot|\alpha)$ is the Dirichlet distribution (A.8). This example will be used for cases where the observations $\{Y_i\}$ are continuous and nonlinear normal.

In the second example, we consider the binomial/beta case (A.5),(A.6):

$$F_Y(Y_i|\theta_i) = Bin(Y_i|\theta_i),$$
$$(\theta|\varphi_k) \sim Beta(\theta|(a_k, b_k), \quad \varphi_k = (a_k, b_k).$$

This example will be used for cases where the observations $\{Y_i\}$ are discrete but the range of the observations is bounded.

In the third example, we consider the Poisson/gamma case (A.3), (A.4):

$$F_Y(Y_i|\theta_i) = Pois(Y_i|\theta_i),$$
$$(\theta|\varphi_k) \sim G(\theta|(a_k, b_k)), \quad \varphi_k = (a_k, b_k).$$

This example will be used for cases where the observations $\{Y_i\}$ are discrete but the range of the observations is unbounded.

In the fourth example, $Y_i \sim F_Y(Y_i|\theta_i)$, $i = 1, ..., N$ and the mixture distribution for θ_i will be of the form

$$(\theta_i|\varphi_k) \sim \sum_{k=1}^{K} w_k \delta_{\varphi_k}(\cdot),$$

where the random vectors $\{\varphi_k\}$ are iid from a known distribution G_0, and where δ_φ is the delta distribution which has mass one at φ. The weights $\{w_k\}$ are defined from the so-called stick-breaking process [Ishwaran and James (2001)] as follows:

$$v_k \sim Beta(\cdot|1, \alpha), \quad k = 1, ..., K - 1; \quad v_K = 1$$
$$w_1 = v_1; \quad w_k = (1 - v_1)(1 - v_2) \cdots (1 - v_{k-1})v_k, \quad k = 2, ..., K,$$

where α is a hyper-parameter which may have a prior distribution. It is straightforward to check that $\sum_{k=1}^{K} w_k = 1$. This example will be used in Chapter 6 to show that the Dirichlet process can be approximated by a finite mixture distribution.

2.2.3 *Normal mixture models*

Assume that we have K distinct sources that correspond to K distinct probability distributions. The parameters θ_i are assumed to come from the following mixture of normals:

$$\theta_i \sim \sum_{k=1}^{K} w_k N(\cdot|\mu_k, \Sigma_k), \qquad (2.24)$$

where w_k are mixture proportions of the model satisfying $\sum_{k=1}^{K} w_k = 1$ and $w_k > 0$ for all k. The distribution of mixture proportions is given by a Dirichlet distribution (A.8) with a density proportional to $w_1^{\alpha_1} \times \cdots \times w_K^{\alpha_K}$, in which case we write:

$$w \sim Dir(\cdot|\alpha), \ w = (w_1, ..., w_K), \ \alpha = (\alpha_1, ..., \alpha_K). \qquad (2.25)$$

If there is no information about the abundance of different sources, then $\alpha_k = 1$, for $k = 1, ..., K$ (non-informative prior). It follows that the nonlinear normal mixture model is:

First stage:

$$Y_i \sim N(\cdot|h_i(\theta_i), \tau^{-1}\Omega_i). \tag{2.26}$$

Second stage:

$$\theta_i \sim \sum_{k=1}^{K} w_k N(\cdot|\mu_k, \Sigma_k). \tag{2.27}$$

Third stage:

$$\tau \sim G(\cdot|a, b),$$
$$\mu_k \sim N(\cdot|\lambda_k, \Lambda_k),$$
$$\Sigma_k^{-1} \sim W(\cdot|(q_k, \Psi_k), \tag{2.28}$$
$$w \sim Dir(\cdot|\alpha),$$

and the hyper-parameters $\{a, b, \alpha_k, \lambda_k, \Lambda_k, q_k, \Psi_k\}$ for $\{k = 1, ..., K, \Omega_i, i = 1, ..., N\}$ are based on prior studies or are chosen to have minimal effect on the analysis. The above Bayesian nonlinear mixture model will be one of our main interests in this book.

From a Bayesian point of view, the analysis of the mixture model given by Eqs. (2.26)–(2.28) requires calculation of the posterior distribution:

$$p(\theta_{1:N}, w_{1:K}, \mu_{1:K}, \Sigma_{1:K}, \tau|Y_{1:N}), \tag{2.29}$$

where for any sequence of vectors $\{v_1, v_2, ..., v_\kappa\}$, we denote $v_{1:\kappa} = \{v_1, v_2, ..., v_\kappa\}$. Generally, this is a very hard problem; most of the statistical literature deals with very special linear cases of Eqs. (2.26)–(2.28). We discuss this in more detail in what follows.

2.2.4 *Nonlinear normal mixture models and pharmacokinetic examples*

Pharmacokinetics is a science that studies the process of absorbtion, distribution and elimination of drugs in an organism. Deep understanding of these processes and the ability to develop personalized approaches to drug treatments is essential in prevention of adverse drug reactions. Adverse drug reaction (ADR) is defined as "an appreciably harmful or unpleasant reaction, resulting from an intervention related to the use of a medicinal product, which predicts hazard from future administration and warrants prevention or specific treatment, or alteration of the dosage regimen, or withdrawal of the product" [Edwards and Aronson (2000)]. In 2004, in a study conducted by the researchers from the University of Liverpool, it was determined that a cause of hospital admission of 1,225 out of 18,820 adult patients was related to ADR; the overall fatality rate was 0.15% [Pirmohamed *et al.* (2004)]. The annual cost of such admissions was an astonishing 466,000,000 pounds sterling.

During an earlier (1994) study in the US it was found that ADRs accounted for over 2.2 million serious cases and more than 100,000 deaths, making ADRs one of the leading causes of hospitalization and death in US [NCBI (2004b)]. Pharmaceutical companies develop medications aimed at an average patient, but therefore they are not particularly suited for specific individuals. Integration of rich patient-specific genomic and pharmacokinetic information will allow an individualized approach to treatment.

In this section we will describe a mathematical approach to this problem. We propose to use time-series observations for gene expression and pharmacokinetics measurements to partition patients into groups that differ by response to a certain drug. Participating individuals can be genotyped and genetic differences between groups can be analyzed. These differences are likely to be associated with differences in drug metabolism. Upon validation, genotyping of patients and subsequent assignment to a drug metabolism group can become an initial stage of any drug treatment.

For the purpose of mathematical modeling, the organism is presented as a set of "compartments" that contain portions of the administered drug. The simplest of all is the one-compartment model. In this case, the decline of the concentration C of the drug in the organism t hours after the administration of the drug can be described by the exponential function

$$C(t) = \frac{D}{V} exp(-\kappa t), \qquad (2.30)$$

where D is the administered dosage, κ is the patient-specific elimination constant, and V is a volume of distribution that is specific to the drug as well as to the patient. Drugs that are highly lipid soluble have a very high volume of distribution and those that are lipid insoluble remain in the blood and have a low V.

If we consider administering a drug to a group of N subjects and taking a series of T measurements of the drug level in the blood, the measured levels can be described by the following PK model:

$$Y_{ij} = \frac{D}{V_i} exp(-\kappa_i t_j) + \sigma_e e_{ij}, \qquad (2.31)$$

where Y_{ij} is a measurement for the ith subject at time t_j, for $i = 1, ..., N$ and $j = 1, ..., T$. Further, measurement errors e_{ij} are assumed to have independent normal $N(0, 1)$ distributions. The patient-specific elimination constants κ_i can be described as a mixture of K normal distributions:

$$\kappa_i \sim \sum_{k=1}^{K} w_k N(\cdot | \mu_k, \sigma_k^2). \qquad (2.32)$$

The patient-specific volume variables V_i can be described by a single normal distribution $V_i \sim N(\cdot | V_0, \sigma_v{}^2)$. For $K = 2$, this reflects the nature of fast and slow acetylators (fast and slow acetylators differ in inherited (genetic) ability to metabolize certain drugs). Therefore, the resulting model is as follows:

$$Y_{ij} = \frac{D}{V_i} exp(-\kappa_i t_j) + \sigma_e e_{ij}, \ \text{for } i = 1, ..., N, \ j = 1, ..., T,$$

$$\kappa_i \sim \sum_{k=1}^{2} w_k N(\cdot | \mu_k, \sigma_k^2),$$

$$\mu_k \sim N(\cdot | \mu_0, \sigma_0^2), \ \text{for } k = 1, ..., K, \qquad (2.33)$$

$$w \sim Dir(\cdot | \alpha), \ \text{where all } \alpha_k = 1,$$

$$V_i \sim N(\cdot | V_0, \sigma_v^2).$$

This model can be used to describe the bimodal nature of certain drug under IV administration.

2.2.5   *Linear normal mixture models and Eyes example*

In the simplest case of $h_i(\theta) \equiv \theta$, the model of Eqs. (2.26)–(2.28) is then reduced to the *linear mixture model*:

$$Y_i \sim N(\cdot|\theta_i, \tau^{-1}\Omega_i),$$

$$\theta_i \sim \sum_{k=1}^{K} w_k N(\cdot|\mu_k, \Sigma_k),$$

$$\tau \sim G(\cdot|a, b,) \qquad\qquad (2.34)$$

$$\mu_k \sim N(\cdot|\lambda_k, \Lambda_k),$$

$$\Sigma_k^{-1} \sim W(\cdot|q_k, \Psi_k),$$

$$w \sim Dir(\cdot|\alpha).$$

We will illustrate this linear normal mixture model with the Eyes example from the WinBUGS manual [Spiegelhalter *et al.* (2003)]. This example was originally described by Bowmaker *et al.* (1985), who presented data on the peak sensitivity wavelengths for individual microspectrophotometric records on a small set of monkeys' eyes. This example has been studied in a number of other papers and books, for example Congdon (2001). In this special (but not uncommon) case $Y_i \equiv \theta_i$, $\sigma^2 = 0$ (see Eq. (2.35)). The wavelength measurements $Y_i, i = 1, ..., 48$ for one monkey are as follows:

> 529.0 530.0 532.0 533.1 533.4 533.6 533.7 534.1 534.8 535.3
>
> 535.4 535.9 536.1 536.3 536.4 536.6 537.0 537.4 537.5 538.3
>
> 538.5 538.6 539.4 539.6 540.4 540.8 542.0 542.8 543.0 543.5
>
> 543.8 543.9 545.3 546.2 548.8 548.7 548.9 549.0 549.4 549.9
>
> 550.6 551.2 551.4 551.5 551.6 552.8 552.9 553.2.

In the histogram in Figure 2.1, the distribution of peak sensitivity wavelength of monkey eyes measured by microspectrophotometry is bimodal and we have to use the mixture model with $K = 2$ components to describe it. We assume that sensitivity wavelength observations $\{Y_i\}$ originate from one of the two groups and can be described by a mixture of normal distributions with the same variance. An unknown fraction of observations w_1 belongs to the first group and $w_2 = 1 - w_1$ belongs to the second group. The equations below describe the famous *Eyes* model with independent

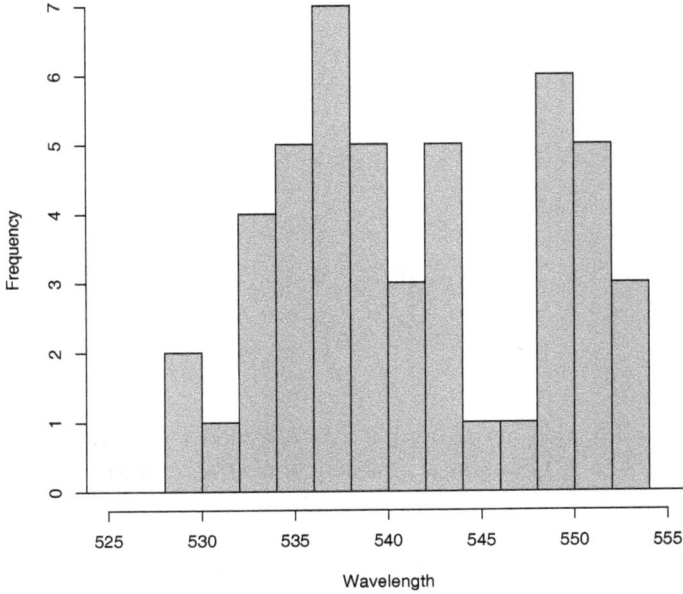

Fig. 2.1. Distribution of peak sensitivity wavelength (nm) of monkey eyes measured by microspectrophotometry.

"non-informative" priors:

$$Y_i \equiv \theta_i,$$

$$\theta_i \sim w_1 N(\cdot|\mu_1, \sigma^2) + w_2 N(\cdot|\mu_2, \sigma^2)$$

$$w \sim Dir(\cdot|\alpha),$$

$$\sigma^{-2} \sim G(\cdot|a, b), \qquad\qquad (2.35)$$

$$\mu_1 \sim N(\cdot|\lambda_1, \sigma_1^2),\ \mu_2 \sim N(\cdot|\lambda_2, \sigma_2^2),$$

$$\lambda_1 = 540,\ \lambda_2 = 540,$$

$$\sigma_1 = \sigma_2 = 10^3.$$

In this book we will use this simple model (Eqs. (2.35)) to illustrate the main methods and concepts of mixture models in general.

2.2.6 *Allocation formulation*

It is beneficial from the computational point of view to choose a different representation of a general mixture model described in Eqs. (2.26)–(2.28). Instead of the traditional formulation:

$$\theta_i \sim \sum_{k=1}^{K} w_k N(\cdot|\mu_k, \Sigma_k),$$
$$w \sim Dir(\cdot|\alpha), \tag{2.36}$$

one can use an alternative:

$$\theta_i \sim N(\cdot|\mu_{Z_i}, \Sigma_{Z_i}), \tag{2.37}$$

where Z_i, $i \in \{1, ..., N\}$ is a discrete random variable taking values in the set $\{1, ..., K\}$ such that $P(Z_i = j) = w_j$. We write this as

$$Z_i \sim Cat(w),$$
$$w \sim Dir(\cdot|\alpha), \tag{2.38}$$

where $Cat(\cdot)$ is a categorical distribution, described in the Eq. (A.9). The $\{Z_i\}$ are called the *allocation*, *latent* or *hidden* variables. They play an important part in many Bayesian and maximum likelihood algorithms and expectation-maximization algorithms [Robert and Casella (2004)].

The allocation parametrization is more convenient for the Gibbs sampling, since one has to sample from one of the normal distributions with a probability w_k, rather than from the sum of many normals. This formulation is actually closer to the real-life case, where observations come from one of the "bins" with some probability associated with each bin. In terms of the general nonlinear normal mixture model of Eqs. (2.26)–(2.28), the allocation formulation becomes:

First stage (unchanged):

$$Y_i \sim N(\cdot|h_i(\theta_i), \tau^{-1}\Omega_i). \tag{2.39}$$

Second stage:

$$\theta_i \sim N(\cdot|\mu_{Z_i}, \Sigma_{Z_i}). \tag{2.40}$$

Third stage:

$$Z_i \sim Cat(w),$$
$$\tau \sim G(\cdot|a, b),$$
$$\mu_k \sim N(\cdot|\lambda_k, \Lambda_k),$$
$$\{\Sigma_k\}^{-1} \sim W(\cdot|q_k, \Psi_k),$$
$$w \sim Dir(\cdot|\alpha_1, ..., \alpha_K). \tag{2.41}$$

2.3   Gibbs Sampling

Let $\Phi = (w_{1:K}, \mu_{1:K}, \Sigma_{1:K}, \tau)$ be the natural parameters of the model of Eqs. (2.39)–(2.41) where again for any vector $\nu = (\nu_1, \nu_2, ..., \nu_m)$ we write $\nu = \nu_{1:m}$. The standard Bayesian estimation problem is to calculate the conditional density $P(\Phi|Y_{1:N})$. From Bayes' theorem we have $P(\Phi|Y_{1:N}) = C \times P(Y_{1:N}|\Phi)P(\Phi)$, where $P(Y_{1:N}|\Phi)$ is the likelihood of Φ, $P(\Phi)$ is the prior density and $C^{-1} = p(Y_{1:N})$. For nonlinear models (i.e. $h_i(\theta_i)$ is a nonlinear function of θ_i) the calculation of the likelihood,

$$P(Y_{1:N}|\Phi) = \prod_{i=1}^{N}\sum_{k=1}^{K} w_k \int N(Y_i|h_i(\theta_i), \tau^{-1}\Omega_i)N(\theta_i|\mu_k, \Sigma_k)d\theta_i, \quad (2.42)$$

is problematic since the integral must be evaluated numerically or by sampling methods, see Wang *et al.* (2007, 2009). For most algorithms, this calculation must be done many times and becomes a serious deterrent to the efficiency of the algorithm. For linear models, (i.e. $h_i(\theta_i)$ is an affine function of θ_i) there is no problem as the likelihood can be evaluated analytically.

Consequently, an algorithm that avoids the calculation of the likelihood is much desired for nonlinear models. The Gibbs sampler is such an algorithm since the likelihood is never calculated. Instead, the parameter space is enlarged to include $\theta_{1:N}$ and $Z_{1:N}$.

Let $\{\alpha_{1:M}\}$ be the collection of all parameters $\theta_{1:N}$, $Z_{1:N}$, $w_{1:K}$, $\mu_{1:K}$, $\Sigma_{1:K}$, and τ of the model. The Gibbs sampler requires calculation of the full conditional distributions:

$$p(\alpha_j|\alpha_{-j}, Y_{1:N}), \; \alpha_{-j} = (\alpha_1, ..., \alpha_{j-1}, \alpha_{j+1}, ..., \alpha_M), \; j = 1, ..., M. \quad (2.43)$$

Fortunately, for models of the type Eqs. (2.39)–(2.41), this is possible [Wakefield *et al.* (1994)]. As described earlier, the Bayesian estimation approach is to calculate the conditional distribution of all the unknown variables given the data, i.e.

$$p(Z_{1:N}, \theta_{1:N}, w_{1:K}, \mu_{1:K}, \Sigma_{1:K}, \tau|Y_{1:N}). \quad (2.44)$$

The Gibbs sampler algorithm outlined in Section 2.1 can be used for this purpose. Since the Gibbs sampler works by iteratively sampling from the full conditional distributions, it requires knowledge of full conditional distributions of model parameters [Gilks *et al.* (1996)]. In the equations below, the notation "..." means conditioning on the data and all remaining variables, e.g.

$$\theta_i|... = \theta_i|\theta_{-i}, Y_{1:N}, w_{1:K}, \mu_{1:K}, \Sigma_{1:K}, \tau. \tag{2.45}$$

For the nonlinear normal mixture model of Eqs. (2.39)–(2.41), the full conditional distributions are given by:

$$(w_1, ..., w_K)|... \sim Dir(\cdot|\alpha_1 + n_1, ..., \alpha_K + n_K), \text{where } n_k = \sum_{i=1, Z_i=k}^{N} 1,$$

$$\mu_k|... \sim N(\cdot|\lambda_k^*, \Lambda_k^*), \text{where } \Lambda_k^* = (n_k \Sigma_k^{-1} + \Lambda_k^{-1})^{-1},$$

$$\lambda_k^* = \Lambda_k(n_k \Sigma_k^{-1} \theta_k^* + \Lambda_k^{-1} \lambda_k), \text{ and } \theta_k^* = \frac{1}{n_k} \sum_{i=1, Z_i=k}^{N} \theta_i,$$

$$(\Sigma_k)^{-1}|... \sim W(\cdot|n_k + q_k, (\Psi_k^{-1} + A_k)^{-1}),$$

$$\text{where } A_k = \sum_{i:Z_i=k} (\theta_i - \mu_k)(\theta_i - \mu_k)^T,$$

$$\tau|... \sim G(\cdot|a + \frac{M}{2}, b + \frac{D}{2}), \text{ where } M = \sum_{i=1}^{N} N_i, N_i = dim(Y_i),$$

$$D = \sum_{i=1}^{N} (Y_i - h_i(\theta_i))^T \Omega_i^{-1} (Y_i - h_i(\theta_i)),$$

$$p(Z_i = k|\theta_i, \mu_{1:K}, \Sigma_{1:K}, w_{1:K}) = \frac{w_k p(\theta_i|\mu_k, \Sigma_k)}{\sum_{j=1}^{K} w_j p(\theta_i|\mu_j, \Sigma_j)},$$

$$\theta_i|... \sim c \times p(Y_i|\theta_i, \tau) p(\theta_i|\mu_{Z_i}, \Sigma_{Z_i}), \tag{2.46}$$

where c is a normalization constant.

Sampling from all the above conditional distributions is straightforward except for $\theta_i|....$ In the nonlinear case this requires a Metropolis–Hastings sampler. For this purpose we have investigated a number of possible sampling variations (some recommended in Bennett *et al.* (1996)) and have adopted a "random-walk" sampler [Wakefield *et al.* (1994)].

In the special case, where there is only one component in the mixture, i.e. $K = 1$, the above results are in accordance with the expressions presented by Wakefield *et al.* (1994). Details of the computation are given in Appendix B.

2.3.1 *Theoretical convergence of the Gibbs sampler*

Assume that all the conditional distributions in Eq. (2.46) are strictly positive on their respective domains. Then the resulting Gibbs sampler is irreducible and aperiodic. Consequently, the two "ergodic" theorems given in Eqs. (2.2) and (2.3) hold. The proof of this theorem was published by Diebolt and Robert (1994) and was proven in an earlier technical report [Diebolt and Robert (1990)]. From Eq. (2.46), we see that the full conditionals for all the random variables except $\{\theta_i\}$ have known distributions. The full conditional distribution of θ_i is only known up to a "normalizing constant". There are several methods for sampling from such distributions: rejection sampling, ratio-of-uniforms, adaptive rejection sampling, Metropolis, slice sampling and other hybrid methods. All these methods are suitable for sampling from general probability densities because they do not require evaluation of their normalizing constants. Bennett *et al.* (1996) evaluated a number of sampling methods in Bayesian estimation for a single component model used in population pharmacokinetics.

Consider the hybrid Gibbs–Metropolis sampling method that is commonly used to sample from a conditional distribution of θ_i. Putting together various sources and ideas, it can be demonstrated that the resulting hybrid Gibbs–Metropolis chain is irreducible and aperiodic. In this section, we just touch on the main ideas of the proof.

Notice that the hybrid Gibbs–Metropolis algorithm with the full conditional distributions given by Eq. (2.46) is a special case of Algorithm A.43 from Robert and Casella (2004, p. 393) when conditional densities q_i are assumed to be symmetric. Robert and Casella (2004, p. 393) state that the hybrid Gibbs–Metropolis algorithm generates a Markov chain with the same invariant distribution as the original Gibbs chain; the proof of this theorem is left for students as an exercise. In a 1990 technical report, Mueller (1990) proved that a "pure" Metropolis chain is irreducible and aperiodic. He also suggested how to extend this result to a hybrid Gibbs–Metropolis chain. In the next section, we illustrate the result for the special case of one Gibbs step and one Metropolis step. The general case deals with n Gibbs steps and m Metropolis steps; some of the Gibbs steps deal with discrete random variable $\{Z_i\}$, while the rest deal with continuous random variables.

2.3.2 Irreducibility and aperiodicity of the hybrid Gibbs–Metropolis chain

Let (X_1, X_2) be a pair of random vectors of a space $E = E_1 \times E_2$ with probability density $\pi(X_1, X_2)$, where E_1 and E_2 are measurable subspaces of Euclidian space. Let $Y_2 \in E_2$ be a random vector and let $q(Y_2|X_2)$ be a symmetric conditional density, i.e., $q(Y_2|X_2) = q(X_2|Y_2)$. Assume that $\pi(X_1, X_2) > 0$ for all $(X_1, X_2) \in E$ and $q(Y_2|X_2) > 0$ for all $Y_2, X_2 \in E_2$, and π is absolutely continuous with respect to the Lebesgue measure. Define the hybrid Gibbs–Metropolis algorithm by Algorithm 2.2.

Algorithm 2.2 Hybrid Gibbs–Metropolis algorithm

Ensure: 1: Gibbs step

Given (X_1^t, X_2^t) simulate $Y_1^{t+1} \sim \pi(Y_1^{t+1}|X_2^t)$.

Ensure: 2: Metropolis step

Simulate $\tilde{Y}_2 \sim q(\tilde{Y}_2|X_2^t)$.

Ensure: 3: Take

$$Y_2^{t+1} = \begin{cases} \tilde{Y}_2, & \text{with probability } \rho; \\ X_2^t, & \text{with probability } 1 - \rho, \end{cases}$$

where $\rho = 1 \wedge \frac{\pi(\tilde{Y}_2|Y_1^{t+1})}{\pi(X_2^t|Y_1^{t+1})}$.

Ensure: 4: Set $(X_1^{t+1}, X_2^{t+1}) = (Y_1^{t+1}, Y_2^{t+1})$, increment $t = t + 1$ and **go to** Step 1.

Algorithm 2.2 is a special case of Algorithm A.43 from Robert and Casella (2004, p. 393), by taking $i = 2$ and $q_2(\cdot)$ to be symmetric.

Theorem 2.3. *The above assumptions and resulting Algorithm 2.2 define a Markov chain $\{X_1^t, X_2^t\}$ with invariant density $\pi(X_1, X_2)$ which is irreducible and aperiodic. Therefore, Eqs. (2.2) and (2.3) hold.*

Proof. The Markov property and invariant density have already been discussed; therefore, we first need to show that the chain is irreducible. For simplicity, write $X = (X_1, X_2)$ and $Y = (Y_1, Y_2)$. It is easy to see that the transition probability for the chain is

$$P(Y \in A|X) \equiv T(X, A) = q(X)1_A(X) + \int_A \rho(X, Y)G(X, Y)dY, \quad (2.47)$$

where $A \subset E$ and $1_A(X)$ is the indicator function of A,

$$q(X) = 1 - \int_A \rho(X,Y)G(X,Y)dY$$

(2.48)

and $G(X,Y) = q(Y_2|X_2)\pi(Y_1|X_2)$.

The next step is to show a condition that implies that the chain is irreducible. By definition of irreducibility, if $\pi(A) > 0$ for some $A \subset E$, then $P(X^t \in A|X^0 = x) > 0$ for all $x \in E$. We show that this is true for $t = 1$. Let $Y = X^1$ and $X = X^0$. We have

$$P(Y \in A|X) \geq \int_A \rho(X,Y)G(X,Y)dY = \int_{(A \cap B) \cup (A \cap B^c)} \rho(X,Y)G(X,Y)dY.$$

(2.49)

Denote

$$I_1 = \int_{A \cap B} \rho(X,Y)G(X,Y)dY$$

(2.50)

and $I_2 = \int_{A \cap B^c} \rho(X,Y)G(X,Y)dY.$

Now fix $x_2 \in E_2$ and define $B(x_2) = B = \{Y = (Y_1,Y_2) : \pi(Y_2|Y_2) < \pi(X_2|Y_1)\}$; so that $\rho(X,Y) = \frac{\pi(Y_2|Y_1)}{\pi(X_2|Y_1)}$ on B. Therefore, on B,

$$\rho(X,Y)G(X,Y) = \frac{\pi(Y_2|Y_1)}{\pi(X_2|Y_1)}q(Y_2|X_2)\pi(Y_1|X_2).$$

(2.51)

Using the definition of conditional probability:

$$\rho(X,Y)G(X,Y) = \frac{\frac{\pi(Y_1,Y_2)}{\pi(Y_1)}}{\frac{\pi(Y_1,X_2)}{\pi(Y_1)}}q(Y_2|X_2)\frac{\pi(Y_1,X_2)}{\pi(X_2)}$$

(2.52)

and canceling out terms $\pi(Y_1)$ and $\pi(Y_1,X_2)$ in numerator and denominator:

$$\rho(X,Y)G(X,Y) = \frac{\pi(Y_1,Y_2)}{\pi(X_2)}q(Y_2|X_2).$$

(2.53)

By definitions of B and ρ, on B^c $\rho(X,Y) = 1 \wedge \frac{\pi(Y_2|Y_1)}{\pi(X_2|Y_1)} \equiv 1$. Therefore, integrals I_1 and I_2 can be written as

$$I_1 = \int_{A \cap B} \rho(X, Y)G(X, Y)dY = \int_{A \cap B} \frac{\pi(Y_1, Y_2)}{\pi(X_2)} q(Y_2|X_2)dY_1dY_2$$

$$= \frac{\int_{A \cap B} \pi(Y_1, Y_2)q(Y_2|X_2)dY_1dY_2}{\pi(X_2)},$$

(2.54)

$$I_2 = \int_{A \cap B^c} \rho(X, Y)G(X, Y)dY = \int_{A \cap B^c} G(X, Y)dY$$

$$= \int_{A \cap B^c} \pi(Y_1|X_2)q(Y_2|X_2)dY_1dY_2 = \frac{\int_{A \cap B^c} \pi(Y_1, X_2)q(Y_2|X_2)dY_1dY_2}{\pi(X_2)}.$$

To prove irreducibility we have to show that $P(Y \in A|X) > 0$ for $A \subset E$ and for all X. Since $P(Y \in A|X) \geq \int_A \rho(X, Y)G(X, Y)dY = I_1 + I_2$ and $I_{1,2} \geq 0$, it is sufficient to show that either $I_1 > 0$ or $I_2 > 0$. We have assumed that $\pi(X_1, X_2) > 0$ for all $(X_1, X_2) \in E$ and $q(Y_2|X) > 0$ for all $\{Y_2, X_2\} \in E_2$, and π is absolutely continuous with respect to Lebesgue measure μ. To prove that either $I_1 > 0$ or $I_2 > 0$, assume the contrary, that both $I_1 = 0$ and $I_2 = 0$ for some X. It means that for the Lebesgue measure $\mu(A \cap B) = 0$ and $\mu(A \cap B^c) = 0$. Since $A = (A \cap B) \bigcup (A \cap B^c)$, then $\mu(A) = \mu(A \cap B) + \mu(A \cap B^c) = 0$. By absolute continuity of measure μ, $\pi(A) = 0$. That contradicts the assumption that $\pi(X_1, X_2) > 0$ for all $(X_1, X_2) \in E$. Hence we have proven the irreducibility of our Markov chain. The rest of the proof follows the steps outlined by Mueller (1990). Note that in Mueller (1990), a stronger result than what we have stated is proven, namely that the Markov chain is "Harris recurrent". This concept is very important in MCMC theory and is named after the late University of Southern California mathematics professor Ted Harris. $\square$

2.3.3 *WinBUGS and JAGS*

In this book, we use the WinBUGS and JAGS statistical software. WinBUGS is an interactive Windows program for Bayesian analysis of complex statistical models using MCMC techniques. WinBUGS is an abbreviation of Windows-based Bayesian inference using Gibbs sampling. WinBUGS was developed by the MRC Biostatistics Unit, at the University of Cambridge, United Kingdom. It is based on the BUGS (Bayesian inference using Gibbs sampling) project started in 1989. The project was established and maintained by Spiegelhalter *et al.* (2006, 2011). WinBUGS provides an

Table 2.1   WinBUGS sampling methods.

Continuous	
Conjugate	Direct sampling using standard algorithms
Log-concave	Derivative-free adaptive rejection sampling
Restricted range	Slice sampling
Unrestricted range	Current point Metropolis

Discrete	
Finite upper bound	Inversion
Shifted Poisson	Direct sampling using standard algorithm

efficient implementation of the Gibbs sampler, with an expert system to select the best sampling method for a given model. The WinBUGS software has been instrumental in raising awareness of Bayesian modeling among both academic and commercial communities, and has enjoyed considerable success over its 20-year life span [Lunn *et al.* (2009)].

One of the best features of WinBUGS is that the user does not need to have special programming skills: the model is written in almost "plain mathematical language". Here is an example from the WinBUGS manual [Spiegelhalter *et al.* (2003)]: if we need to say that $Z_i \sim Cat(\cdot|w)$ and $y_i \sim N(\cdot|\mu_{Z_i}, \tau^{-1})$, this can be written in the BUGS language as

$$
\begin{aligned}
&\text{for (i in 1: N) \{} \\
&\qquad \text{Z[i] } \sim \text{ dcat(w[])} \\
&\qquad \text{y[i] } \sim \text{ dnorm(mu[Z[i]], tau)\}}
\end{aligned}
\tag{2.55}
$$

The user does not have to specify full-conditional distributions – they are computed by WinBUGS. In addition, WinBUGS offers a convenient and visual method of model formulation: the Doodle editor. An example of the Doodle for a three-component mixture model is shown in Figure 2.2. Doodle operates with three kinds of objects: nodes, plates and edges. Nodes correspond to either stochastic and logical variables or constants. Edges connecting nodes represent functional relationships between variables. Plates are used to represent arrays of variables and observations.

The sampling methods used in WinBUGS correspond to the hierarchies listed in the Table 2.1. In each case, a method is only used if no previous method in the hierarchy is appropriate [Spiegelhalter *et al.* (2003)].

The choice of a sampling method determines the required minimum lengths of the Markov chain. For example, the sampling method required for the full conditional distribution of θ_i in Eq. (2.46) is called the *current point Metropolis* algorithm. This method is based on a univariate independent symmetric random walk with proposal density $N(\cdot|\kappa, \sigma^2)$, where κ is

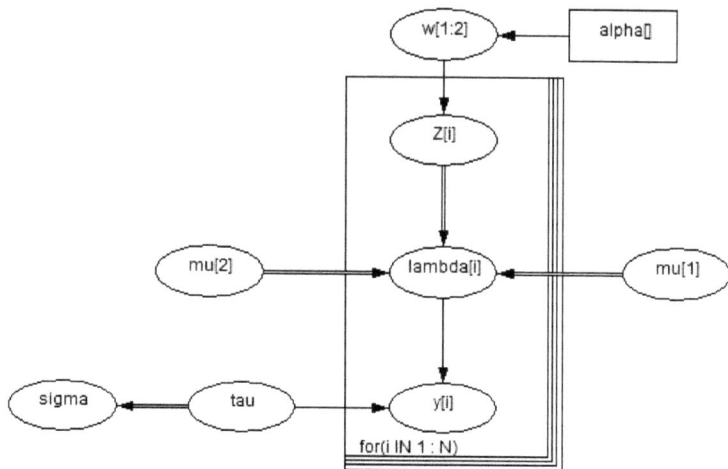

Fig. 2.2.   An example of model formulation in Doodle editor: nodes correspond to either stochastic and logical variables or constants; edges connecting nodes represent functional relationships between variables; plates are used to represent arrays of variables and observations.

the current value of the component in question and σ is a scale factor. This method uses the first 4,000 iterations for tuning of the scale factor σ to get an acceptance rate between 20% and 40%. The current point Metropolis is used by WinBUGS for unrestricted non-log-concave distributions [Spiegelhalter *et al.* (2003); Gelman and Rubin (1992)].

Unfortunately, it is now time to mention a popular Russian saying: "free cheese exists only in the mouse trap." WinBUGS users do not have to supply the full conditional distribution, select the sampling order in the Gibbs sampler, specify the Metropolis proposal density, and sample from exotic distributions such as Wishart. However, the simplicity of WinBUGS usage comes with the restriction that the user is not allowed to perform certain calculations inside the program between the steps of the Gibbs sampler. For example, calculations for the random permutation sampler (Section 3.3), KLMCMC (Section 4.4) and BDMCMC (Section 3.5) algorithms must be done outside of WinBUGS. In general, the kind of calculations that are excluded are those that change the state of the Markov chain. This leaves a choice to the user: either to code the Gibbs sampler from scratch and include the necessary calculations or to interrupt the execution of WinBUGS, perform the calculations and restart.

In a recent critical assessment of WinBUGS, Lunn *et al.* (2009) pointed out that in WinBUGS there is no mechanism in place to ensure that a given model is identifiable. Also, Lunn *et al.* (2009) found that the BUGS language does not discriminate between models for which the software is good at fitting and those for which it may perform poorly. In addition, when there are multiple ways to specify a model, they may lead to dramatically different behaviors and posteriors.

Plummer (2003) developed an efficient alternative to WinBUGS, "Just Another Gibbs Sampler" (JAGS). JAGS is platform independent. Pre-processing and post-processing of data, model formulation, and results interpretation are done outside of JAGS. The package RJags offers a convenient way of communication between JAGS and the statistical language *R* [Hornik (2013)]. The program does not have a graphical user interface, but it is suitable for analyzing large tasks using high-performance computing clusters. JAGS syntax is slightly different from that of WinBUGS; it is more amenable to user-extension. In the JAGS implementation several flaws in the BUGS language were corrected; for example, JAGS achieves better separation of modeling assumptions from data [Lunn *et al.* (2009)].

2.4   Prior Distributions: Linear and Nonlinear Cases

Clever selection of the prior distributions and parameters frequently determines the success of the entire modeling process. Conjugate priors, although sometimes too restrictive, ensure the manageable form of the full conditional distributions and ease of sampling. Unfortunately, it is not always possible to describe the problem in such regular terms. For the subset of linear problems, the parameters of the priors can be estimated using sample means and sample variance.

Nonlinear problems require more careful development and individual approaches. Sometimes, past datasets can provide the range for the parameters. When *a priori* information is not available, it is common practice to use non-informative or vaguely informative priors. For some problems this choice can, however, lead to the instability of the simulated chains and improper posterior distributions [Carlin (1999)]. When variances of the prior distribution are selected to be too small (as will be illustrated in Eq. (3.23) below), the Gibbs sampler might get "trapped" in one of the labeling subspaces. This, in turn, results in the incorrect estimation of the model parameters. Below, we will present examples of selection of prior parameters for several linear and nonlinear problems.

Analysis of the selection process of prior distributions shows how mixture models differ from non-mixture models. For example, for the linear mixture model (Eq. (2.34)), if we set $K = 1$, then the prior distributions can be chosen to be non-informative and the resulting posterior distributions will still be proper. However, this is not the case for $K \geq 2$: if the $\{\phi_k\}$ are independent and the prior distributions are non-informative, then the resulting posterior distributions will be improper. This was demonstrated by Marin *et al.* (2005) in a paper on computational difficulties with mixture models. A brief proof is given below.

Consider the general linear mixture model:

$$\theta \sim \sum_{k=1}^{K} w_k f_k(\cdot|\phi_k).$$

Denote by $\phi = (\phi_1, ..., \phi_K)$, $w = (w_1, ..., w_K)$, and $Y = (Y_1, ..., Y_N)$. Let the components of ϕ be independent and let ϕ and w be independent. Then the prior distribution of (ϕ, w) is: $\pi(\phi, w) = \pi(\phi_1) \cdots \pi(\phi_K)\pi(w)$.

Now assume that the prior of ϕ_k is improper for all k, i.e. $\int d\phi \pi_k(\phi_k) = \infty$, for all k. Using

$$p(\phi, w|Y) \propto f(Y|\phi)\pi(\phi, w), \tag{2.56}$$

the posterior distribution is of the form

$$p(\phi, w|Y) \propto \left(\prod_{i=1}^{N} \sum_{k=1}^{K} w_k f(Y_i|\phi_k) \right) \pi(\phi, w). \tag{2.57}$$

It follows:

$$\int \int d\phi dw p(\phi, w|Y) \propto \int \int d\phi dw \left(\prod_{i=1}^{N} \sum_{k=1}^{K} (w_k f(Y_i|\phi_k)) \right) \pi(\phi)\pi(w)$$

$$= \int \int d\phi dw_1 \pi(w)(w_1)^N f(Y_1|\phi_1) \cdots f(Y_N|\phi_1)\pi(\phi)$$

$$+ \text{ (other non-negative terms)} \tag{2.58}$$

$$\geq \int \int dw\pi(w)d\phi_1(w_1)^N f(Y_1|\phi_1) \cdots f(Y_N|\phi_1)\pi(\phi_1)$$

$$\times \int d\phi_2 \pi(\phi_2) \cdots \int d\phi_K \pi(\phi_k) = \infty.$$

Consequently, only the informative (and, possibly, data-dependent) priors should be used for mixture models. Below, we consider three cases of normal mixture models to illustrate this point:

(1) $p(Y_i) = \sum_{k=1}^{K} w_k N(Y_i|\mu_k, \sigma_k^2)$

(2) $p(Y_i) = \sum_{k=1}^{K} w_k N(Y_i|\mu_k, \Sigma_k)$, where μ_k is a vector and Σ_k is a variance–covariance matrix

(3) $p(Y_i) = N(Y_i|h_i(\theta_i), \Sigma_0)$ and $p(\theta_i) = \sum_{k=1}^{K} w_k N(\theta_i|\mu_k, \sigma_k^2)$, where h_i is some nonlinear function of the parameter θ

The first two cases are by far the most common models in the literature; just a few references exist for Case (3). In the first case, the observations come from normal one-dimensional mixture distributions and thus the prior distributions can be selected as

$$\pi(\mu_k) = N(\mu_k|\mu_0, s^2),$$

$$\pi(\sigma_k^{-2}) = G(\sigma_k^{-2}|\alpha, \beta),$$

$$(2.59)$$

where μ_0, s, α and β are some fixed hyper-parameters. The common practice is to choose $\mu_0 = mean\{Y_{1:N}\}$ and $s = range\{Y_{1:N}\}$. The choice of α and β is not that straightforward and there are several ways to select them. In WinBUGS examples, both α and β are always chosen to be 0.001. This is a conveniently wide and flat distribution with a variance of $1,000$ and mean 1. Another choice of hyper-parameters was proposed by Richardson and Green (1997). They suggested using a fixed α and putting a gamma distribution on β. Although some studies (e.g. [Stephens (1997b)]) indicate that the fixed and variable β have the same effect on the MCMC performance, the choice of variable β has become increasingly popular among statisticians.

In the second case, observations $Y_i = \{y_{i1}, ..., y_{iq}\}$ and parameters μ_k are vectors in q-dimensional space and Σ_k are matrices. The priors are usually chosen as

$$\mu_k \sim N(\cdot|\mu_0, S),$$

$$\Sigma_k^{-1} \sim W(\cdot|(R\rho)^{-1}, \rho).$$

$$(2.60)$$

As before, $\mu_0 = mean\{Y_{1:N}\}$ and $S = diag[r_1, r_2, ..., r_q]$, where r_q is the range of $\{y_{iq}\}$. The hyper-parameter ρ is usually selected to be equal to the dimension q of the vectors. Sometimes, "vague" priors are placed on S and R:

$$S^{-1} \sim W(\cdot|(S^0 a)^{-1}, a) \text{ and } R \sim W(\cdot|(R^0 b)^{-1}, b), \qquad (2.61)$$

where a and b are equal to the dimension of matrices and the suitable values S^0 and R^0 are selected based on the fact that $E(R) = R^0$ and $E(S) = S^0$.

The selection of priors for nonlinear mixture models did not get much attention in the literature, possibly due to the difficulty level of the issue. It seems feasible to perform a preliminary model fitting for $K = 1$ and estimate posterior means $\bar{\theta}_i$ of θ_i. Then use $\bar{\theta}_i$ in place of observations and find mean and range. Finally, use these values to construct prior distributions as in Cases (1) and (2). Later in this book we will suggest other methods of prior distribution selection for the nonlinear model.

Chapter 3

Label Switching and Trapping

We start this chapter with a description of the permutation invariance properties of mixture models as well as the label-switching effect (Section 3.1). Next we address the main convergence issues of Markov chains and the methods of convergence diagnostics (Section 3.2). Then we describe the random permutation sampler algorithm which utilizes the properties of label switching to determine convergence of the Markov chain (Section 3.3). However, label switching does not allow determination of component-specific parameters. Consequently, we then describe algorithms that try to minimize the influence of label switching: methods of re-parametrization (Section 3.4) and relabeling (Section 3.5).

3.1 Label Switching and Permutation Invariance

3.1.1 *Label switching*

Label switching is a term describing the invariance of the likelihood under relabeling of the mixture components, as defined by Stephens (2000b). As a result, model parameters have multimodal symmetric posterior distributions. We will now define these notions in more precise terms.

We consider the hierarchical, possibly nonlinear, mixture model of the form

$$Y_i | \theta_i \sim p_i(\cdot | \theta_i), i = 1, ..., N,$$

$$(3.1)$$

$$\theta_i \sim \sum_{k=1}^{K} w_k f(\cdot | \varphi_k),$$

where the $\{\theta_i\}$ are independent and identically distributed (iid) and the measurements $\{Y_i\}$ are independent but not necessarily identically distributed (inid); $p_i(\cdot|\theta)$ is a probability distribution that may depend on the index i; and $f(\cdot|\varphi)$ is a probability density that does not depend on the index k.

Let $W = (w_1, ..., w_K)$, $\Phi = (\varphi_1, ..., \varphi_K)$, $\Psi = (W, \Phi)$ and $Y = (Y_1, ..., Y_N)$. Then,

$$p(Y_i|\Psi) = \int p_i(Y_i|\theta_i)p(\theta_i|W, \Phi)d\theta_i = \int p_i(Y_i|\theta_i) \sum_{k=1}^{K} w_k f(\theta_i|\varphi_k)d\theta_i$$

$$= \int p_i(Y_i|\theta_i)S(\theta_i|W, \Phi)d\theta_i \quad (3.2)$$

where $S(\theta|\Psi) = \sum_{k=1}^{K} w_k f(\theta|\varphi_k)$ and

$$p(Y|\Psi) = \prod_{i=1}^{N} \int p_i(Y_i|\theta_i)S(\theta_i|\Psi)d\theta_i. \quad (3.3)$$

3.1.2 *Permutation invariance*

Permutation invariance of the mixture distribution has a surprising and very important consequence: all the marginal posterior distributions of the component parameters, component weights and component labels are identical. Let $X = (x_1, ..., x_K)$ be any K-dimensional vector and let ν be any permutation of the set $\{1, ..., K\}$. Then write $\nu(X) = (x_{\nu(1)}, ..., x_{\nu(K)})$.

Definition 3.1. A function $f(\Psi)$ is *permutation invariant* if $f(\Psi) = f(\nu(\Psi)) = f(\nu(W), \nu(\Phi))$ for every permutation ν of the set $\{1, ..., K\}$.

A remarkable property of permutation invariance of $\Psi = (W, \Phi)$ is proven in Theorem 3.1. We first need two lemmas.

Lemma 3.1. *Assume the model of Eq. (3.1). Then the likelihood $p(Y|\Psi)$ as a function of Ψ is permutation invariant for every Y.*

Proof. As a function of $\Psi = (W, \Phi)$, the sum of $S(\theta|W, \Phi) = \sum_{k=1}^{K} w_k f(\theta|\varphi_k)$ is clearly permutation invariant for every θ. Consequently, so are $p_i(Y_i|\Psi)$ for each i and Y_i and $p(Y|\Psi)$ for every Y, since

$$p_i(Y_i|\Psi) = \int p_i(Y_i|\theta_i)S(\theta_i|\Psi)d\theta_i = \int p_i(Y_i|\theta_i)S(\theta_i|\nu(\Psi))d\theta_i$$

$$= p_i(Y_i|\nu(\Psi)). \quad (3.4)$$

Similarly,

$$p(Y|\Psi) = \prod_{i=1}^{N} \int p_i(Y_i|\theta_i)S(\theta_i|\Psi)d\theta_i = \prod_{i=1}^{N} \int p_i(Y_i|\theta_i)S(\theta_i|\nu(\Psi))d\theta_i$$

$$= p(Y|\nu(\Psi)). \tag{3.5}$$

Hence, we have demonstrated that $p_i(Y_i|W,\Phi)$ and $p(Y|W,\Phi)$ are permutation invariant, *q.e.d.* □

Lemma 3.2. *Assume the model of Eq. (3.1) and assume the prior density $p(W,\Phi)$ is permutation invariant. Then the posterior distribution $p(W,\Phi|Y)$ is permutation invariant for every Y.*

Proof. From Bayes' rule, $p(\Psi|Y) = \frac{p(Y|\Psi)p(\Psi)}{p(Y)}$, so that $p(\nu(\Psi)|Y) = \frac{p(Y|\nu(\Psi))p(\nu(\Psi))}{p(Y)} = p(\Psi|Y)$ from Lemma 3.1.

□

The permutation invariance of the posterior density $p(\Psi|Y)$ has some surprising and important results for the marginal posteriors $p(\varphi_k|Y)$ for $k = 1, ..., K$: they are all identical. The same is true for the marginal posteriors $p(w_k|Y)$ for $k = 1, ..., K$.

Theorem 3.1. *Assume the model of Eq. (3.1) and assume the prior density $p(\Psi)$ is permutation invariant. Then*

$$p(\varphi_k|Y) = p(\varphi_j|Y),$$

$$\tag{3.6}$$

$$p(w_k|Y) = p(w_j|Y)$$

for all $k = 1, ..., K$, and $j = 1, ..., K$.

Proof. Marginalizing $p(\Phi, W|Y)$ over all variables in (Φ, W) except φ_k we have

$$p(\varphi_k|Y) = \int p(\Phi, W|Y)d(\varphi_1, ..., \varphi_{k-1}, \varphi_{k+1}, ..., \varphi_K)d(w_1, ..., w_K). \tag{3.7}$$

Now change the variables of integration with the transform $\Phi \to \nu(\Phi), W \to \nu(W)$ where ν is any permutation of the set $\{1, ..., K\}$. The Jacobean of this transformation is 1. It follows that

$$p(\varphi_k|Y) = \int p(\nu(\Phi), \nu(W)|Y)d(\varphi_{\nu(1)}, ..., \varphi_{\nu(k-1)}, \varphi_{\nu(k+1)}, ..., \varphi_{\nu(K)})$$

$$\times \ d(w_{\nu(1)}, ..., w_{\nu(K)}). \tag{3.8}$$

By Lemma 3.1,

$$p(\varphi_k|Y) = \int p(\Phi, W)|Y)d(\varphi_{\nu(1)}, ..., \varphi_{\nu(k-1)}, \varphi_{\nu(k+1)}, ..., \varphi_{\nu(K)})$$

$$\times d(w_{\nu(1)}, ..., w_{\nu(K)}). \tag{3.9}$$

But now the integration in Eq. (3.9) is marginalization with respect to all the variables except $\varphi_{\nu(k)}$, so the right-hand side of Eq. (3.9) is $p(\varphi_{\nu(k)}|Y)$. Therefore $p(\varphi_k|Y) = p(\varphi_{\nu(k)}|Y)$. Since the permutation ν was arbitrary, we have proved the first equation in Eq. (3.6); a similar argument proves the second equation in Eq. (3.6), *q.e.d.*. □

The results in the above lemmas and theorem are proved by Fruhwirth-Schnatter (2006) for the case of iid measurements and a mixture model of the form: $Y_i \sim \sum_{k=1}^{K} w_k f(\cdot|\varphi_k)$. Our results extend this to the hierarchical inid case.

3.1.3   *Allocation variables* $\{Z_i\}$

The permutation invariance of $p(\Psi|Y)$ also has a surprising effect on the allocation variables $\{Z_i\}$. To prove permutation invariance for the allocation variables $\{Z_i\}$, we need to consider a simpler mixture model, namely,

$$Y_i \sim \sum_{k=1}^{K} w_k f(\cdot|\varphi_k), \tag{3.10}$$

where now the $\{Y_i\}$ are iid. This can be obtained from our general hierarchical mixture model of Eq. (3.10) by setting $Y_i = \theta_i$, $i = 1, ..., N$. The model of Eq. (3.10) is exactly the same as that considered in Chapter 3 of Fruhwirth-Schnatter (2006). In this case, the allocation variables $\{Z_i\}$ are defined by: $P(Z_i = k) = w_k$ and $p(Y_i|Z_i = k) = f(Y_i|\varphi_k)$.

Theorem 3.2. *Assume the model of Eq. (3.10) and assume the prior density $p(\Psi)$ is permutation invariant. Then $p(Z_i = k|Y) = p(Z_i = j|Y)$ for all Y, $k = 1, ..., K, j = 1, ..., K$ and $i = 1, ..., N$.*

Proof. First, note

$$P(Z_i = k|W, \Phi, Y) = \frac{P(Y|Z_i = k, W, \Phi)P(Z_i = k|W, \Phi)}{P(Y|W, \Phi)}$$

$$= \frac{\prod_{j=1, j\neq i}^{N} P(Y_j|W, \Phi)P(Y_i|Z_i = k, W, \Phi)w_k}{\prod_{j=1, j\neq i}^{N} P(Y_j|W, \Phi)P(Y_i|W, \Phi)}$$

$$= \frac{P(Y_i|Z_i = k, W, \Phi)w_k}{P(Y_i|W, \Phi)} = \frac{f(Y_i|\varphi_k)w_k}{P(Y_i|W, \Phi)}. \tag{3.11}$$

Then,

$$p(Z_i = k|Y) = \int p(Z_i = k, W, \Phi|Y)dWd\Phi$$

$$= \int p(Z_i = k|W, \Phi, Y)p(W, \Phi|Y)dWd\Phi$$

$$= \int \frac{f(Y_i|\varphi_k)w_k}{P(Y_i|W, \Phi)}p(W, \Phi|Y)dWd\Phi. \qquad (3.12)$$

Now change the variables of integration with the transform $\Phi \to \nu(\Phi)$, $W \to \nu(W)$ where ν is any permutation of the set $\{1, ..., K\}$. The Jacobean of this transformation is equal to 1. Then,

$$p(Z_i = k|Y) = \int \frac{f(Y_i|\varphi_{v(k)})w_{v(k)}}{P(Y_i|\nu(W), \nu(\Phi))}p(\nu(W), \nu(\Phi)|Y)d\nu(W)d\nu\Phi). \quad (3.13)$$

By Theorem 3.1, $P(Y_i|W, \Phi) = P(Y_i|\nu(W), \nu(\Phi))$ and $p(W, \Phi|Y) = P(\nu(W), \nu(\Phi)|Y)$. It follows that

$$p(Z_i = k|Y) = \int \frac{f(Y_i|\varphi_{v(k)})w_{v(k)}}{P(Y_i|W, \Phi)}p(W, \Phi|Y)dWd\Phi. \qquad (3.14)$$

But the right-hand side of this last equation equals $p(Z_i = v(k)|Y)$. Therefore $p(Z_i = k|Y) = p(Z_i = v(k)|Y) = p(Z_i = j|Y)$, for any $j = 1, ..., K$, *q.e.d.*. □

As a corollary of Theorem 3.2, an even more surprising result is obtained:

Corollary 3.1. *Assume the model of Eq. (3.10) and assume the prior density $p(\Psi)$ is permutation invariant. Then $P(Z_i = k|Y) = 1/K$ for all Y, $k = 1, ..., K$ and $i = 1, ..., N$.*

Proof. We have $\sum_{k=1}^{K} p(Z_i = k|Y) = 1$. From Theorem 3.2, all the terms in the sum are equal. Therefore $P(Z_i = k|Y) = 1/K$., *q.e.d.* □

3.1.4 *An example of label switching*

The label switching property of mixture models can be illustrated by the normal mixture example (Eyes) taken from the WinBUGS 1.4 manual [Spiegelhalter *et al.* (2003)]. In this case, the model is given by Eq. (2.35) with the initial condition for $\mu = (540, 540)$. Code and data are in Appendix D.

Table 3.1   Eyes example. Results of Gibbs sampler illustrating the switching effect. Sample size is $10,000$, first $1,000$ iterations discarded.

Node	Mean	StDev	MC error	2.5%	Median	97.5%
w_1	0.42	0.11	0.01	0.24	0.41	0.68
w_2	0.58	0.11	0.01	0.32	0.59	0.76
μ_1	547.6	3.79	0.605	536.2	548.7	551
μ_2	537.9	3.71	0.604	535	536.8	549.7
τ	0.07	0.02	4.7×10^{-4}	0.032	0.07	0.12

Using the initial conditions of Eq. (2.35), we ran this model in Win-BUGS, discarding the first 1,000 burn-in iterations and sampling from the last 10,000 iterations. We observed the effects of some label switching as shown in Figures 3.1, 3.2 and Table 3.1. Figure 3.1 shows that the label-switching is not complete and the Gibbs sampler has not converged. We can also see that the posterior distribution for μ is bimodal. There are $K! = 2$ possible ways of labeling the components; thus, the posterior distributions of the model parameters can have up to 2! modes.

3.2   Markov Chain Convergence

Analysis of convergence consists of two major parts. First, one must determine a moment where the Markov chain "forgets" about its starting point, or, more precisely, when the "values simulated from a given Markov chain can be considered to be realizations from its stationary distribution" [Lunn *et al.* (2002)]. Second, determine how many samples must one generate in order to obtain accurate inference about model parameters.

Raftery and Lewis

Gilks *et al.* (1996, Chapter 7) proposed a numeric algorithm to find the minimum number of iterations to achieve the required estimation precision for some function of model parameters U. Assume that one needs to estimate the posterior probability $P(U \leq u | Data)$ to within $\pm r$ with probability s, when the quantile of interest is q. They found that the required number of iterations N_{min} is

$$N_{min} = \left\{ F^{-1}\left(\frac{s+1}{2}\right) \right\}^2 \frac{q(1-q)}{r^2}, \tag{3.15}$$

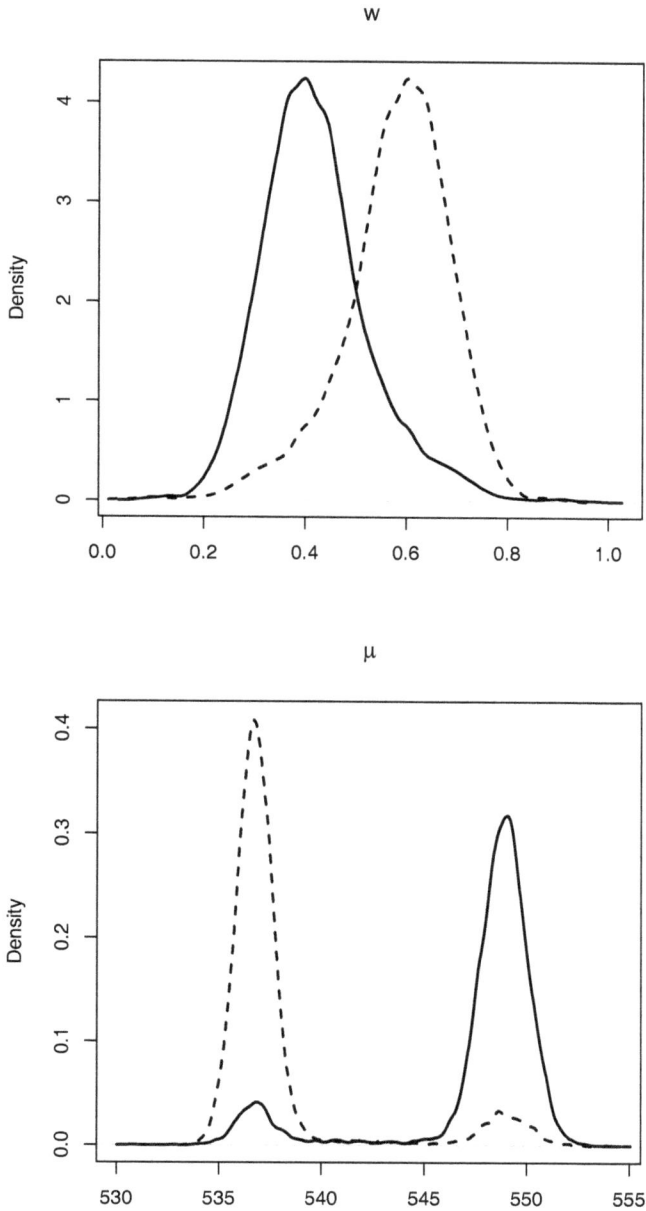

Fig. 3.1. Eyes example. Simulated posterior densities for component weights w and means μ. Solid lines μ_1 and w_1; dotted lines μ_2 and w_2.

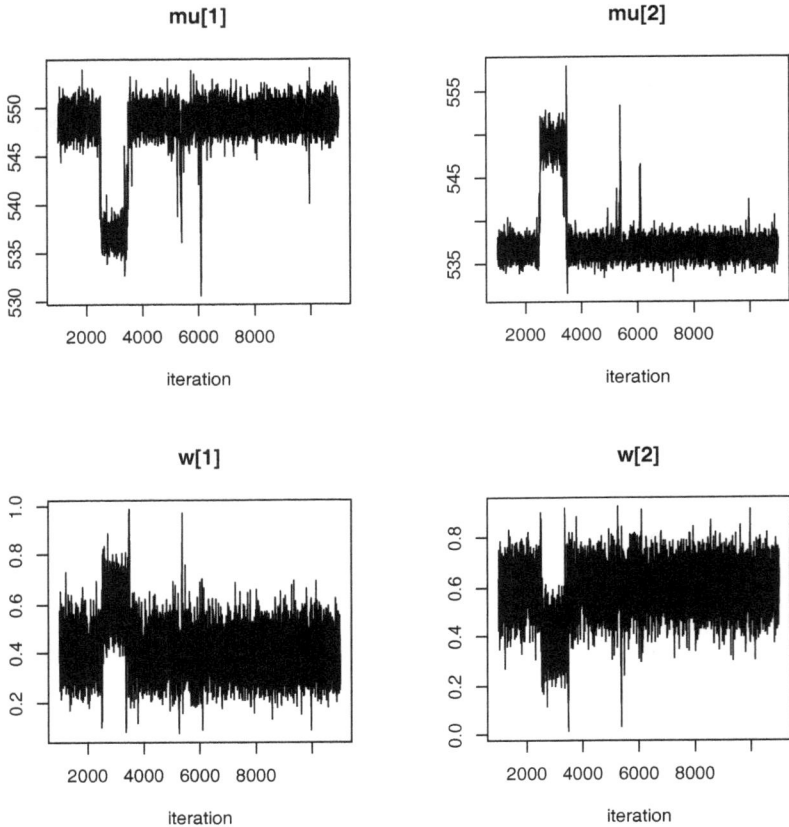

Fig. 3.2.   Eyes example. History plots for component weights w and means μ.

where $F(\cdot)$ is the standard normal cumulative distribution function. The dependence factor for the Raftery and Lewis convergence diagnostics measures the multiplicative increase in the number of iterations needed to reach convergence due to within-chain correlation [Smith (2005)]. It was observed by Raftery and Lewis (1992) that dependence factors greater than 5.0 indicate convergence failure.

Gelman and Rubin

It is convenient to estimate the convergence in the framework of WinBUGS, since it allows us to run the simulation of multiple chains with different starting points. Just by visually inspecting history plots we can see how close the various chains are. Additionally, if we download the CODA output for a parameter of interest ϱ, we can utilize, for instance, the "industry standard" approach outlined by Gelman and Rubin (1992) and Gilks *et al.* (1996, Chapter 8). For m parallel simulations, let us label parallel sequences of length n as ϱ_{ij}, $i = 1, ..., m$, $j = 1, ..., n$ and compute two values: B (between-sequence variance) and W (within-sequence variance).

$$B = \frac{n}{m-1} \sum_{i=1}^{m} (\bar{\varrho}_{i\cdot} - \bar{\varrho}_{\cdot\cdot})^2, \text{ where } \bar{\varrho}_{i\cdot} = \frac{1}{n} \sum_{j=1}^{n} \varrho_{ij} \text{ and } \bar{\varrho}_{\cdot\cdot} = \frac{1}{m} \sum_{i=1}^{m} \bar{\varrho}_{i\cdot}$$

$$(3.16)$$

$$W = \frac{1}{m} \sum_{i=1}^{m} s_i^2, \text{ where } s_i^2 = \frac{1}{n-1} \sum_{j=1}^{n} (\varrho_{ij} - \bar{\varrho}_{i\cdot})^2.$$

A conservative estimate of the variance of ϱ under over-dispersion is then constructed from B and W:

$$\widehat{var(\varrho)} = \frac{n-1}{n} W + \frac{1}{n} B.$$

$$(3.17)$$

Since W and $\widehat{var(\varrho)}$ are lower and upper bounds for a $var(\varrho)$, and both of them converge to $var(\varrho)$ as $n \to \infty$, their ratio $\widehat{R}$, called *the potential scale reduction factor*, can be used as a measure of convergence:

$$\sqrt{\widehat{R}} = \sqrt{\frac{\widehat{var(\varrho)}}{W}}.$$

$$(3.18)$$

For a converged simulation, the potential scale regression factor is expected to be $\sqrt{\widehat{R}} < 1.2$.

Brooks and Gelman

Brooks and Gelman (1998) have proposed an extension of the univariate approach of Gelman and Rubin. They have created a method of convergence diagnostics based on the entire vector of parameters Φ. Between-sequence variance B and within-sequence variance W in the multivariate case are defined as follows:

$$B(\Phi) = \frac{n}{m-1} \sum_{i=1}^{m} (\bar{\Phi}_{i\cdot} - \bar{\Phi}_{\cdot\cdot})(\bar{\Phi}_{i\cdot} - \bar{\Phi}_{\cdot\cdot})^T, \text{ where}$$

$$\bar{\Phi}_{i\cdot} = \frac{1}{n} \sum_{j=1}^{n} \Phi_{ij} \text{ and } \bar{\Phi}_{\cdot\cdot} = \frac{1}{m} \sum_{i=1}^{m} \bar{\Phi}_{i\cdot},$$

$$\qquad (3.19)$$

$$W(\Phi) = \frac{1}{m} \sum_{i=1}^{m} s_i^2, \text{ where } s_i^2 = \frac{1}{n-1} \sum_{j=1}^{n} (\Phi_{ij} - \bar{\theta}_{i\cdot})(\Phi_{ij} - \bar{\Phi}_{i\cdot})^T,$$

$$\widehat{var(\Phi)} = \frac{n-1}{n} W\left(\Phi\right) + \left(1 + \frac{1}{m}\right) \frac{1}{n} B.$$

The *multivariate potential scale reduction factor* $\widehat{MR(\Phi)}$ is defined as a maximum root-statistic measure of distance between $\widehat{var(\Phi)}$ and $W(\Phi)$. For the p-dimensional vector Φ, it is defined by the formula

$$\widehat{MR(\Phi)} = max_{a \in \Re^p} \frac{a^T \widehat{var(\Phi)} a}{a^T W(\Phi) a}. \qquad (3.20)$$

The background diagram function of WinBUGS performs the convergence assessment using the approach of Gelman and Rubin, modified by Brooks and Gelman.

Geweke

Geweke's approach to assessment convergence is based on standard time-series methods. His method is applicable when simulation is performed with only one chain. Geweke suggested dividing the chain into two "windows", containing the first $p\%$ and the last $q\%$ of simulated values, and computing means in both of them:

$$Z = \frac{\bar{X}_q - \bar{X}_p}{\sigma_{(X_q - X_p)}}, \qquad (3.21)$$

where $\sigma_{(X_q - X_p)}$ is an estimated standard error of the difference $X_q - X_p$. If the sampling distribution of $Z \to N(0,1)$, when the length of the chain goes to infinity, then the chain has converged. If the computed Z-score is too big, one needs to discard the first $p\%$ of iterations and re-compute the convergence score.

Autocorrelation

Another possible way to assess the convergence of the Markov chain is by computing the autocorrelations with various lags within each chain. The "ideal" trace plot of the Markov chain should look like a "fat hairy caterpillar". If the chain's convergence and mixing are slow, then it will look like a slow snake on the trace plot, and the autocorrelation will be high. WinBUGS has a built-in functionality to compute the autocorrelation as a function of a lag. Lunn *et al.* (2002) pointed out that a "snake"-like chain requires longer running time; after indefinitely many iterations, a "snake" turns into a "fat hairy caterpillar".

Convergence

Let us illustrate the assessment of convergence using the μ parameter from the Eyes example in Eq. (2.35). We ran the Gibbs sampler with $m = 2$ chains for $N = 4,000$ iterations, discarding the first 2,000 iterations, resulting in the chain length $n = 2,000$. The difference between two chains was in the initial values of $\mu_{1,2}$: it was equal to 535 for the first chain and 545 for the second chain. Using CODA output, we have computed the univariate potential scale regression factor for μ_1; it was equal to $\widehat{R} = 1.08 < 1.2$, and therefore indicated convergence of the analyzed Markov chain. Visual inspection of the plots on Figure 3.3 for μ_1 and μ_2 agrees with our conclusion that the convergence has been achieved.

The mixture model framework provides one more powerful way to assess convergence: via equality of the multimodal posterior distributions. For a K-component mixture model:

$$Y_i \sim \sum_{k=1}^{K} w_k f(\cdot | \phi_k), \tag{3.22}$$

where (w_k, ϕ_k) are component weights and component-specific parameters, and the prior distributions are permutation invariant, then the posterior distributions of parameters should be the same for all components, as shown in Section 3.1. Large differences between posterior distributions for different components indicate that the parameter space was not entirely traversed and the Gibbs sampler got stuck in one of the modes. In the next chapter, we will present our version of the trans-dimensional approach, developed by Richardson and Green (1997) and later by Stephens (2000b), which leads to "improved mixing over model parameters".

A

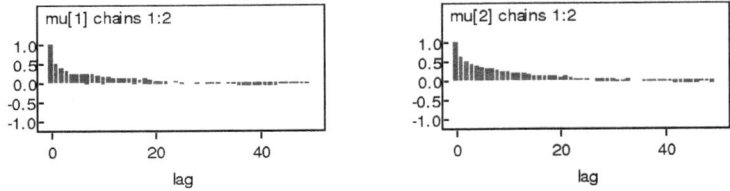

B

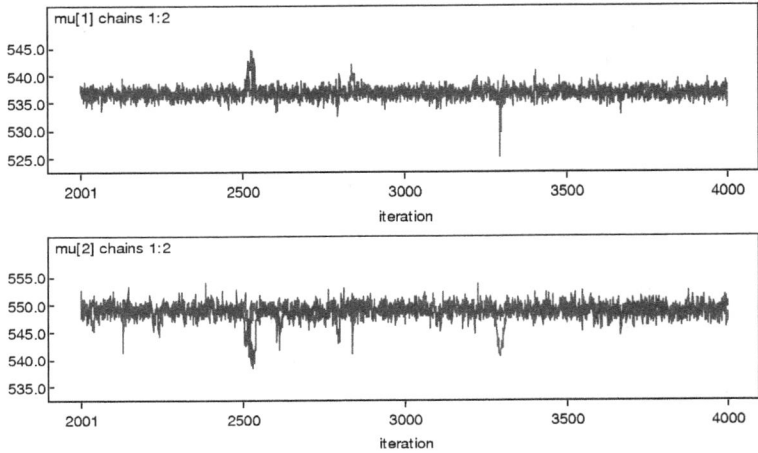

C

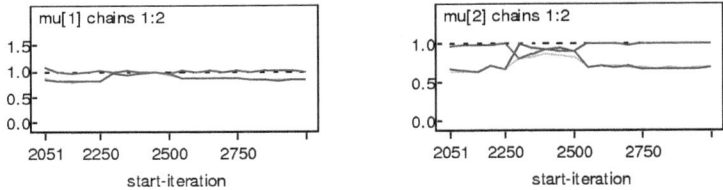

Fig. 3.3. Conversion assessment for the μ parameter from the Eyes example. (A) Autocorrelation plots; (B) Trace plots; (C) Gelman–Rubin statistic.

3.2.1 *Trapping states and convergence*

As we will see later, practical convergence of the Gibbs sampler is a difficult property to achieve. A large number of research papers have been written on the topic of computational difficulties with implementation of mixture models, especially on the poor convergence of the Gibbs sampler, regardless of its theoretical convergence. As we have demonstrated in the previous section, the necessary condition for convergence is the permutation invariance of the posterior distributions and the resulting equality of the various conditional marginal densities. In this regard, a striking statement was made in Celeux *et al.* (2000): "almost the entirety of the Markov chain Monte Carlo (MCMC) samples implemented for mixture models has failed to converge!" They demonstrated this effect with two simple examples (a sum of three univariate Gaussians and a sum of three univariate exponentials) and showed that the standard Gibbs sampler and a random walk Metropolis–Hastings algorithm do not converge based on this necessary permutation invariance. Celeux *et al.* proposed to use the *simulated tempering* due to Neal (1996) that "encourages moves between the different modes in full generality". The *simulated tempering* was ultimately shown to "converge" based on the necessary permutation invariance property. Unfortunately, this algorithm is complex and "entails considerable computational burden". In a later section of this work, we will present a considerably simpler trans-dimensional Gibbs–Metropolis sampler which satisfies the sample "convergence" properties.

Trapping states. What is wrong in the examples of Celeux *et al.* (2000) and in many other examples is that the MCMC chain gets "trapped" in one of the $K!$ modal regions of the posterior surface. Although theoretically the MCMC chain is irreducible and therefore must explore all points of the posterior surface (infinitely often), in practice this is not the case. The Gibbs sampler stays trapped in the current labeling subspace with rare unbalanced jumps to other subspaces. This property of the Gibbs sampler leads to poor estimates of model parameters. As noted by Casella *et al.* (2004), "Even though the Gibbs chain ... is irreducible, the practical setting is one of an almost-absorbing state which is called a *trapping state* as it may require an enormous number of iterations to escape from this state. In extreme cases, the probability of escape is below the minimal precision of the computer and the trapping state is truly absorbing, due to computer rounding errors". There have been many explanations for this trapping phenomenon:

- "The Gibbs sampler becomes effectively trapped when one of the components of the mixture is allocated very few observations" [Chapter 24 of Gilks *et al.* (1996)].
- "This problem can be linked with a potential difficulty of this modeling, namely, that it does not allow a noninformative (or improper) Bayesian approach. Moreover, vague informative priors often have

Table 3.2   Eyes example. Gibbs sampler trapping problem. Sample size $N = 100{,}000$, first 50,000 iterations discarded as burn-in.

Node	Mean	StDev	MC error	2.5%	Median	97.5%
w_1	0.56	0.079	6.0×10^{-5}	0.44	0.60	0.75
w_2	0.44	0.079	6.0×10^{-5}	0.25	0.39	0.56
μ_1	536.6	0.791	0.008	535.0.0	536.7	538.2
μ_2	549.0	1.035	0.011	547.0	549.0	551.0

the effect of increasing occurrences of the trapping states, compared with more informative priors. This is also shown in the lack of proper exploration of the posterior surface, since the Gibbs sampler often exhibits a lack of label-switching" [Casella *et al.* (2004)].

- The Gibbs sampler goes to and subsequently cannot leave one of the many local minima (similar problems exist for the expectation-maximization (EM) algorithm).

This effect can be illustrated by the apparent stickiness of the Gibbs sampler in the case of the Eyes problem Eq. (2.35). Now instead of allowing the precision τ to be random, we will consider a case of known and fixed τ;

$$\begin{cases} Y_i \sim N(\mu_{Z_i}, \sigma^2) \\ Z_i \sim Cat(w) \\ w \sim Dir(\cdot|\alpha), \end{cases} \tag{3.23}$$

where the number of components $K = 2$. We used independent "noninformative" priors:

$$\begin{cases} \alpha = (1,1), \ \tau = 1/\sigma^2, \ \sigma = 3.5 \\ \mu_1 \sim N(540, 10^{-3}), \mu_2 \sim N(540, 10^{-3}). \end{cases} \tag{3.24}$$

Initial conditions were chosen to be $\mu^0 = (540, 540)$.

We ran 150,000 iterations of the Gibbs sampler in WinBUGS, discarding the first 50,000 iterations as burn-in. As we can see from Figure 3.4 and Table 3.2, the Gibbs sampler got trapped at the subspace where $\mu = (536, 549)$, and running the sampler for additional millions of iterations will not change the picture. This is indeed a case of an "absorbing" state. This occurred because of the poor choice of the prior distribution for the parameter σ (it was assumed to be constant). This problem can partially be fixed by applying the birth-death MCMC (BDMCMC) method discussed in the next chapter, and the distribution of parameters becomes bimodal. Due to the stickiness of some Gibbs sampler states, the distributions are

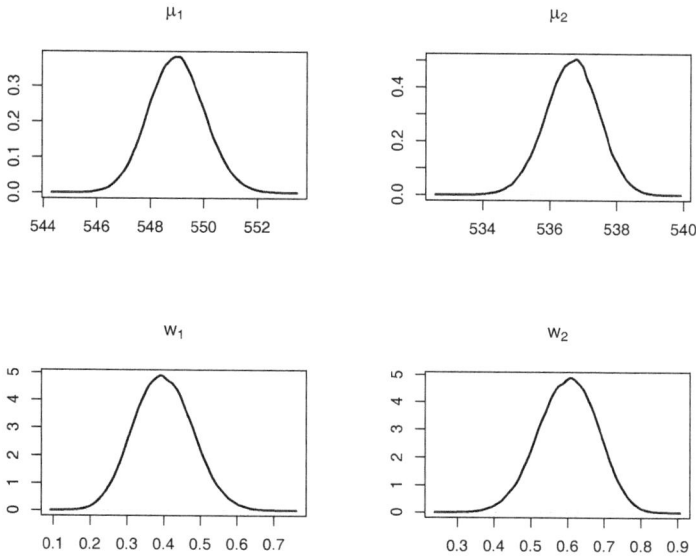

Fig. 3.4. Eyes example. Simulated posterior densities for model parameters μ and w. Gibbs sampler gets trapped in a labeling subspace, leading to poor estimation of model parameters.

close but not identical. The random permutation sampler discussed in the next section may often resolve this problem and make the rarely visited states of the Gibbs sampler more frequent by additionally randomizing the component labels. We must point out that the situation cannot be saved when the state is not visited by the Gibbs sampler at all.

In this example, it is easy to see that convergence has not been achieved by the lack of permutation invariance. If a sampler is allowed to run from two or more sufficiently different points, the chain "converges" to different absorbing states. For example, when we start the Gibbs sample for the Eyes dataset from three sufficiently different points ($\mu^0 = (535, 545)$, $\mu^0 = (545, 535)$ and $\mu^0 = (530, 540)$) the chain "converges" to three different absorbing states. This issue can be easily detected by the Gelman–Rubin convergence criterion (see Figure 3.5). Brooks and Gelman (1998) emphasized that "one should be concerned both with convergence of R to 1, and with convergence of both the pooled and within interval widths to stability". WinBUGS implementation of the Gelman–Rubin test for parameter μ estimates $R \approx 5$, therefore indicating the lack of convergence. Using

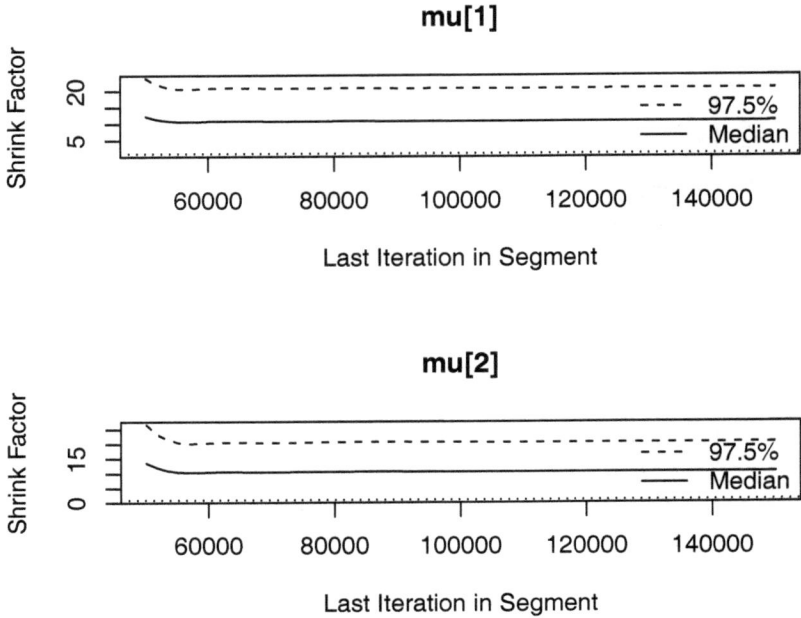

Fig. 3.5.   Eyes example, Gelman-Rubin convergence diagnostics for 3 chains.

Brooks, Gelman and Rubin convergence diagnostic tools in the Bayesian output analysis program (BOA), we estimated the multivariate potential scale reduction factor (Eq. (3.20), Figure 3.7) to be equal to 14.76. Both of these measures indicate that the samples have not arisen from the stationary distribution [Smith (2005)].

We have to point out that other convergence tests fail to detect the lack of convergence. Application of different WinBUGS and BOA convergence tests for two chains starting from two different points ($\mu^0 = (535, 545)$ and $\mu^0 = (545, 535)$) detect the lack of convergence due to the existence of the trapping states. For example, the Raftery and Lewis convergence test (Table 3.3) computes dependence factors for all parameters. The highest dependence factor, calculated for chain 1 for parameter μ_1, is equal to 4.84 < 5.0; it is below the threshold indicating the convergence failure. Heidelberg and Welch's stationarity and interval half-width test did not identify the problem with convergence. The Geweke convergence test

Table 3.3   Eyes example. BOA Raftery and Lewis convergence diagnostic.

Param.	chain	Thin	Burn-in	Total	Lower bound	Dep. factor
μ_1	1	4	8	18160	3746	4.84
μ_2	1	3	6	11907	3746	3.17
μ_1	2	3	6	13686	3746	3.65
μ_2	2	3	6	13311	3746	3.55
μ_1	3	2	4	13701	3746	3.65
μ_2	3	2	4	11730	3746	3.13

Table 3.4   Eyes example. BOA Geweke convergence diagnostic.

Statistic	chain	Z-Score	p-value
μ_1	1	-0.41	0.68
μ_2	1	0.24	0.88
μ_1	2	1.13	0.26
μ_2	2	0.93	0.35
μ_1	3	-0.70	0.48
μ_2	3	-0.67	0.50

(Eq. (3.21), Table 3.4, Figure 3.6) found that for the three chains the computed Z-scores were close to 0 and p-values were > 0.05, and thus there is no evidence against the convergence [Smith (2005)]. Only the Gelman–Rubin and Brooks–Gelman convergence tests were able to identify the problem (Figures 3.5 and 3.7) with the Gibbs sampler output.

3.3   Random Permutation Sampler

Computational difficulties associated with MCMC estimation of mixture models motivated Fruhwirth-Schnatter (2001, 2006) to develop the random permutation sampler (RPS) method. Fruhwirth-Schnatter made a simple but brilliant proposal to facilitate convergence of MCMC samplers for mixture models: conclude each sweep with "a random permutation of the current labeling of the states". Since permutation invariance is a necessary condition for convergence, forcing permutations between the sweeps of the Gibbs sampler cannot violate the properties of the algorithm.

For simplicity, let us describe the RPS algorithm in terms of a general linear mixture model (Eq. (2.34)), where observations $Y = (Y_1, ..., Y_n)$ are assumed to originate from a mixture distribution:

$$Y_i \sim \sum_{k=1}^{K} w_k f(\cdot | \phi_k). \tag{3.25}$$

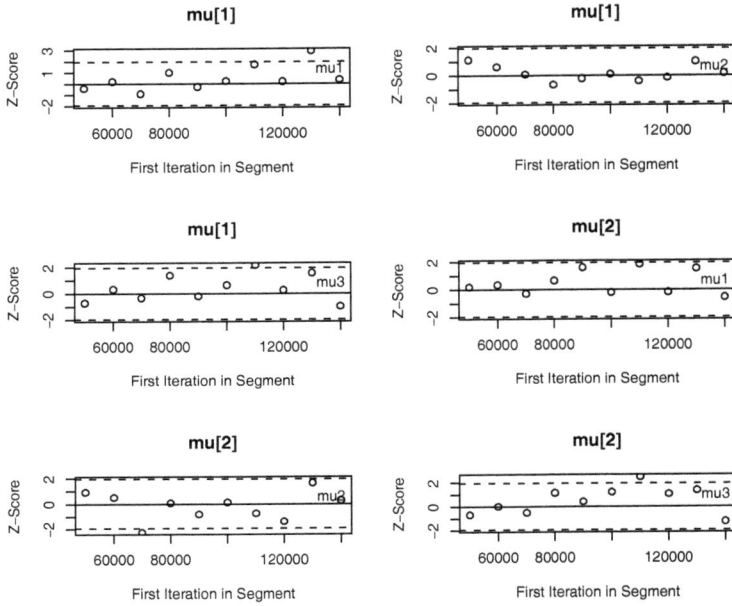

Fig. 3.6.   Eyes example. Geweke convergence diagnostics.

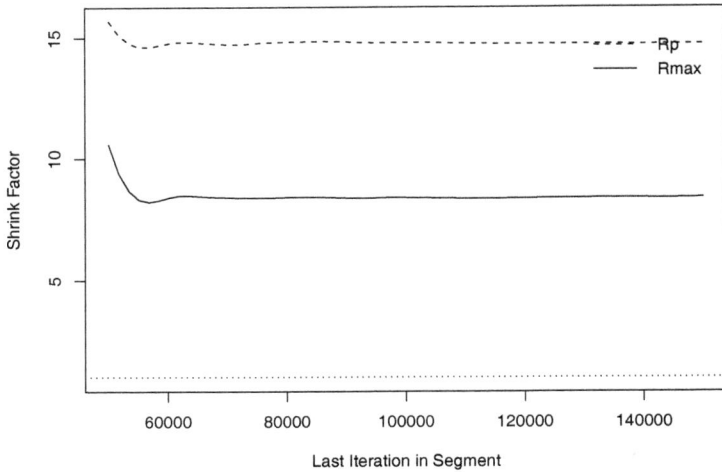

Fig. 3.7.   Eyes example. Convergence diagnostics using Brooks and Gelman multivariate shrink factor.

Let $\pi(\phi, w, Z|Y)$ be the posterior distribution, where ϕ is the set of the distribution parameters, w are the component weights, and Z is the set of the component allocation variables. As before, denote by ν a permutation of the set $\{1, ..., K\}$. The Fruhwirth-Schnatter RPS is given by Algorithm 3.1 below.

Algorithm 3.1 Random permutation sampler

Require: Start with identity permutation ν and some priors for Z, w and ϕ.

Ensure: 1: Generate Z, w and ϕ from $\pi(\phi, w, Z|Y)$.

Ensure: 2: Generate a random permutation ν of the current labeling of the states and define parameters $\tilde{Z} = Z_\nu$, $\tilde{w} = w_\nu$ and $\tilde{\phi} = \phi_\nu$.

Ensure: 3: Using $\tilde{Z}$, $\tilde{w}$ and $\tilde{\phi}$ as priors for the Gibbs sampler, repeat **Steps 1–3**.

This method can easily be generalized to the nonlinear mixture model. As shown by Fruhwirth-Schnatter (2001), if the original MCMC algorithm was convergent, then the Algorithm 3.1 (RPS) was also convergent and the target densities of RPS and the original MCMC were identical. In addition to preserving the convergence properties, the RPS enhances mixing of MCMC.

This can be easily shown using the Eyes dataset with fixed σ: the MCMC gets stuck, no switches occur between the labeling subspaces and consequently the distributions of the component weights and means are unimodal. If we apply the RPS to the same dataset, the distributions become bimodal (Figure 3.8). Moreover, in this case the RPS-modified Gibbs sampler converges to the true posterior (permutation invariant) distribution.

In the simple case of a linear Gaussian mixture, the posterior distribution for the model parameters $\mu = (\mu_1, ..., \mu_K)$ and $w = (w_1, ..., w_K)$ can be calculated numerically:

$$p(\phi, w|Y) = cL(Y|\phi, w)p_0(\mu)p_0(w), \qquad (3.26)$$

where c is a normalizing constant. $L(Y|\phi, w)$ is the likelihood, given by the formula

$$L(Y|\phi, w) = \prod_{i=1}^{N} p(y_i|\phi, w) = \prod_{i=1}^{N} \sum_{k=1}^{k} w_k N(y_i - \mu_k, \sigma_k^2). \qquad (3.27)$$

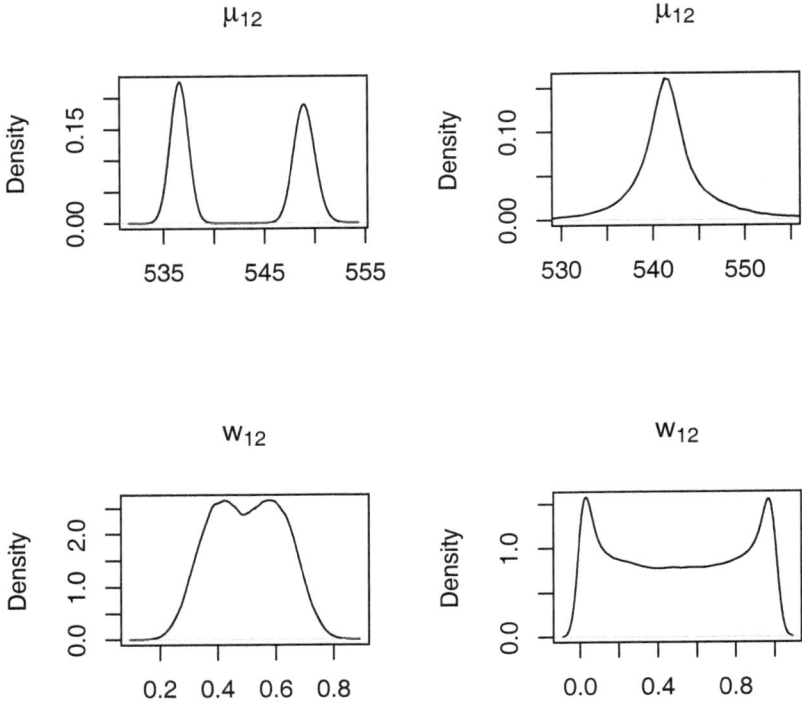

Fig. 3.8. Eyes dataset: $\sigma = 3$ (left) and $\sigma = 7$ (right) posterior distributions of wavelength and component weight as a result of posterior RPS processing.

Prior distributions are: $p_0(\mu_k) = N(\mu_k - \mu_0, \sigma_0^2)$, $p_0(w) = 1$ on $[0, 1]$. Denote $g(\mu, w) \equiv cL(Y|\phi, w)p_0(\mu)p_0(w)$. It follows that

$$p(\mu_k|Y) = c \int \cdots \int d^{K-1}\mu_{m \neq k} \int \cdots \int d^K w_m g(\mu, w),$$

$$p(w|Y) = c \int \cdots \int d^K \mu g(\mu, w). \qquad (3.28)$$

For $K = 2$ and $\sigma_1 = \sigma_2 = \sigma_0 = \sigma$ we calculated the likelihood in Eq. (3.28) and the posterior density in Eq. (3.26) numerically. Varying the values of σ, one can obtain different posterior distributions of μ and w, either bimodal or unimodal (Figure 3.8). However, the RPS and the exact computation always agree.

3.3.1 *RPS post processing*

There are many algorithms that analyze mixture models with an unknown number of components, and we propose one more method in Section 4.4.1 of this work. Each of these algorithms benefits from the RPS modification, and it is surprising that it is not as widely used as it should be. To demonstrate the advantages of using RPS along with the existing methods, it seems to be necessary to do the simulation twice: with and without the RPS modification. However, common sense suggests that if all labeling subspaces are visited by an algorithm, even in an unbalanced fashion, the randomizing step can be done as post-processing on the Gibbs sampler output. The theorems in this section validate this assertion.

Theorem 3.3. *Let*

(1) $G = \{g_1, g_2, ..., g_{K!}\}$ *be the set of all permutations on* $\{1, ..., K\}$; *and let* P *be a probability on* G *such that* $P(g_k) = \alpha_k > 0$ *and* $\sum_{k=1}^{K!} \alpha_k = 1$.

(2) $\phi = \{\phi_1, \phi_2, ..., \phi_K\}$ *be a* K-*dimensional random vector with probability density* $p(\phi)$, *such that* $p(\phi) = p(g(\phi))$, *where* $g(\phi) = (\phi_{g_1}, \phi_{g_2}, ..., \phi_{g_K})$ *for all* $g \in G$.

(3) $F(\phi)$ *be a real-valued* L_1 *function relative to* $p(\phi)$, *i.e.* $E[\|F(\phi)\|] < \infty$,

then $E_{g,\phi}(F(g(\phi))) = E_\phi(F(\phi))$.

Proof. For any $g \in G$, the expected value

$$E_\phi(F(g(\phi))) = \int d\phi p(\phi) F(g(\phi)).$$

Introducing the transformation $\tilde{\phi} = g(\phi)$ and $\phi = g^{-1}(\tilde{\phi})$ with Jacobian $J = 1$:

$$E_\phi(F(g(\phi))) = \int dg^{-1}(\tilde{\phi}) p(\tilde{\phi}) F(\tilde{\phi}) = \int d\phi p(\phi) p(\phi) F(\phi) = E_\phi(F(\phi)). \tag{3.29}$$

Next, notice that $E_{g,\phi}(F(g(\phi))) = \sum_{k=1}^{K!} \alpha_k \int d\phi F(g(\phi)) p(\phi) = \sum_{k=1}^{K} \alpha_k E_\phi(g_k(\phi))$, for $k \in \{1, ..., K\} = E_\phi(F(\phi)) \sum_{k=1}^{K!} \alpha_k = E_\phi(F(\phi))$, q.e.d. □

Theorem 3.4. *Under all assumptions of Theorem 3.3, let* $\{\phi^t\}$ *be a sequence of random vectors, such that for every real-valued function* $F \in L_1$,

$$\frac{\sum_{t=1}^{T} F(\phi^t)}{T} \to E_\phi(F(\phi)) < \infty \text{ a.e. as } T \to \infty. \tag{3.30}$$

If $\{g^t\}$ is a sequence of random permutations uniformly distributed on G, then

$$\frac{\sum_{t=1}^{T} F(g^t(\phi^t))}{T} \rightarrow E_\phi(F(\phi) \text{ a.e. as } T \rightarrow \infty, \tag{3.31}$$

where a.e. means "almost everywhere".

Proof. If $\frac{\sum_{t=1}^{T} F(\phi^t)}{T}$ converges a.e. to $E_\phi(F(\phi))$ as $T \rightarrow \infty$, then for any fixed permutation vector g, the sum $\frac{\sum_{t=1}^{T} F(g(\phi^t))}{T}$ converges a.e. to $E_\phi(F(g(\phi)))$, as $T \rightarrow \infty$. Now consider all $K!$ possible permutations of model parameters. For any fixed T, let

$$S_l = \{t : g^t = g_l, t = 1, ..., T\},$$

$$\tag{3.32}$$

$$T_l = \sharp(S_l)$$

Denote $\alpha_l^T = \frac{T_l}{T}$, for $1 \leq l \leq K!$. Then $0 \leq \alpha_l^T \leq 1$ and $\sum_{l=1}^{K!} \alpha_l^T = 1$. For any $0 < T < \infty$,

$$\frac{\sum_{t=1}^{T} F(g^t(\phi^t))}{T} = \sum_{l=1}^{K!} \frac{\sum_{t \in S_l} F(g_l(\phi^t))}{T_l} \frac{T_l}{T} = \sum_{l=1}^{K!} \alpha_l^T \frac{\sum_{t \in S_l} F(g_l(\phi^t))}{T_l}. \tag{3.33}$$

Define $f_l(T) = \frac{\sum_{t \in S_l} F(g_l(\phi^t))}{T_l}$ and $f = E_\phi(F(\phi))$. Consider the difference

$$\left| \sum_{l}^{K!} \alpha_l^T f_l(T) - f \right| = \left| \sum_{l}^{K!} \alpha_l^T (f_l(T) - f) \right|. \tag{3.34}$$

We have

$$\left| \sum_{l=1}^{K!} \alpha_l^T (f_l(T) - f) \right| \leq \sum_{l=1}^{K!} \alpha_l^T |(f_l(T) - f)|$$

$$\tag{3.35}$$

$$\leq \sum_{l=1}^{K!} \alpha_l^T \times \sum_{l=1}^{K!} |(f_l(T) - f)| = \sum_{l=1}^{K!} |(f_l(T) - f)|.$$

Since $K! < \infty$, we have $T_l \rightarrow \infty$ as $l \rightarrow \infty$. Hence $(f_l(T) - f) \rightarrow 0$ a.e. as

$T \to \infty$. Therefore

$$\sum_{l=1}^{K!} |(f_l(T) - f)| \to 0 \text{ a.e.,}$$

$$\left| \sum_{l=1}^{K!} \alpha_l^T f_l(T) - f \right| \leq \sum_{l=1}^{K!} |f_l(T) - f| \to 0 \text{ a.e. and}$$

$$\sum_{l=1}^{K!} \alpha_l^T \frac{\sum_{t \in t_l} F(g^t{}_l(\phi^t))}{T_l} \to E_\phi(F(\phi)) \text{ a.e.,}$$

$$\text{as } T \to \infty, \ q.e.d.$$

$\square$

Using the RPS as a post-processing step is much easier to implement than embedding it into the existing algorithms. In this work we will use the equivalence of both approaches proven by Theorem 3.4 to make our computation more efficient. Figure 3.8 shows posterior distributions of a peak sensitivity wavelength and a component weight for the Eyes dataset, treated with RPS as post-processing.

3.4 Re-parametrization

Historically, the label-switching property was regarded as a problem and artificial measures were introduced to avoid it. One such measure, called re-parametrization, is the stipulation of identifiability constraints for the model parameters. For example, from the description of the Eyes dataset, we can expect to see two distinct wavelengths $\mu_1 < \mu_2$. This can be programmed by introduction of an additional positive parameter β: $\mu_2 = \mu_1 + \beta$, where $\beta \sim N(0, 10^{-6})$, for $\beta > 0$. Here is the resulting model [Spiegelhalter *et al.* (2003)] with independent "noninformative" priors and re-parametrization:

$$
\begin{aligned}
Y_i &\sim N(\mu_{Z_i}, \sigma^2), \\
Z_i &\sim Cat(P), \\
w &\sim Dir(\cdot | \alpha), \ 1 \leq k \leq K, \\
\tau &= 1/\sigma^2, \ \tau \sim G(10^{-3}, 10^{-3}), \\
\beta &\sim N(0, 10^{-6}), \text{ where } \beta > 0 \\
\mu_1 &\sim N(0, 10^{-6}), \\
\mu_2 &= \mu_1 + \beta \text{ (re-parametrization).}
\end{aligned}
\tag{3.36}
$$

Initial conditions: $\beta = 0$ and $\mu_1 = 540$.

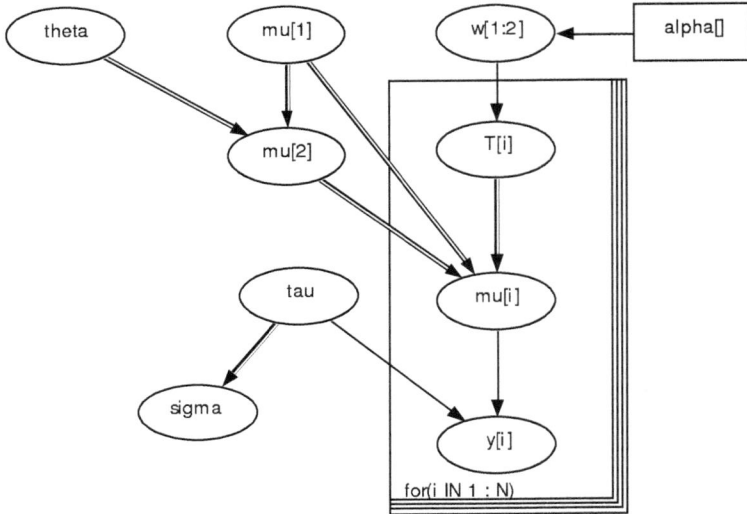

Fig. 3.9. Normal mixture model (with re-parametrization) for peak sensitivity wavelengths for individual microspectrophotometric records on a small set of monkey eyes.

Table 3.5   Eyes example (with re-parametrization). WinBUGS output: Sample size 10,000, first 1,000 iterations discarded as burn-in.

Node	Mean	StDev	MC error	2.5%	Median	97.5%
w_1	0.6	0.087	0.0017	0.42	0.6	0.76
w_2	0.4	0.087	0.0017	0.24	0.4	0.58
μ_1	540.0	0.89	0.019	540.0	540.0	540.0
μ_2	550.0	1.3	0.029	550.0	550.0	550.0
σ	3.8	0.62	0.017	2.9	3.7	5.4

The BUGS code is in Appendix D. A doodle graph of the model is given in the Figure 3.9. After 1,000 burn-in iterations followed by 10,000 iterations, clear separation between two components of the mixture is achieved, as shown in Table 3.5. The posterior densities for w, μ and σ with re-parametrization are unimodal; the wavelength μ_2 has the correct value of 550 and the standard deviation σ with re-parametrization is less than σ without re-parametrization ($\sigma = 3.8$ vs. $\sigma = 4.4$).

Introduction of identifiability constraints removes the label switching in this simple example. However, without prior knowledge of the "true" values of the parameters, we probably would not be able to effectively choose the re-parametrization. In complex datasets, as noted by Richardson and Green (1997), different re-parameterizations will lead to different posterior distributions.

3.5 Stephens' Approach: Relabeling Strategies

Consider again the model of Eq. (3.10),

$$p(Y_i|\Psi) = \sum_{k=1}^{K} w_k f(Y_i|\phi_k), \Psi = (w, \phi),$$

where now the observations $Y_1, ..., Y_N$ are iid and write $Y = (Y_1, ..., Y_N)$.

Estimating parameters, clustering observations into groups and summarizing posterior distributions are all problematic with the label-switching properties of the Markov chain. Imposing artificial identification constraints does not guarantee to correct this label-switching problem. In addition, these constraints do not allow the Gibbs sampler to explore the entire parameter space.

These considerations motivated Stephens to run the unconstrained Gibbs sampler, obtaining possibly $K!$-modal distributions for a K-component mixture, and then perform mixture deconvolution. In a series of publications, Stephens (1997a, 2000b) presented relabeling strategies based on minimizing the posterior expectation of a loss function.

Let the function $L(a; \Psi)$ be the loss due to some action a, when the true value of a model parameter is Ψ, and choose the action $\hat{a}$ to minimize the expected posterior loss $R(a) = E[L(a; \Psi|Y])$. Now assume that $\Psi^{(t)}, t = 1, ..., M$ are sampled values of Ψ (after burn-in) from a Markov chain with stationary distribution $p(\Psi|Y)$. Then the Monte Carlo risk is given approximately by $\hat{R}(a) = (1/M)\sum_{t=1}^{M} L(a; \Psi^{(t)})$. The general algorithm of Stephens is as follows:

Algorithm 3.2 General relabeling algorithm [Stephens (2000b)]

Require: Start with some initial permutations $\nu_1, ..., \nu_M$ (e.g. the identity). Assume that $\Psi^{(t)}, t = 1, ..., M$ are sampled values of Ψ from a Markov chain with stationary distribution $p(\Psi|Y)$. Then iterate the following steps until a fixed point is reached.

Ensure: 1: Choose $\hat{a}$ to minimize $\sum_{t=1}^{M} L(a; \nu_t(\Psi^{(t)}))$

Ensure: 2: For $t = 1, ..., M$, choose ν_t to minimize $L(\hat{a}; \nu_t(\Psi^{(t)}))$

Algorithm 3.3 Classification relabeling algorithm [Stephens (2000b)]

Require: Start with some initial permutations $\nu_1, ..., \nu_M$ (e.g. the identity). Assume that $\Psi^{(t)}, t = 1, ..., M$ are sampled values of Ψ from a Markov chain with stationary distribution $p(\Psi|Y)$. Then iterate the following steps until a fixed point is reached.

Ensure: 1: Choose $\hat{q}_{ij}$ to minimize

$$D = \sum_{t=1}^{M} \sum_{i=1}^{N} \sum_{j=1}^{K} p_{ij}(\nu_t(\Psi^{(t)})) log(p_{ij}(\nu_t(\Psi^{(t)}))/q_{ij}) \tag{3.37}$$

and to satisfy $\sum_{j=1}^{K} \hat{q}_{ij} = 1$, for every i.

Ensure: 2: For $t = 1, ..., M$, choose ν_t to minimize

$$D_t = \sum_{i=1}^{N} \sum_{j=1}^{K} p_{ij}(\nu_t(\Psi^{(t)})) log \frac{p_{ij}(\nu_t(\Psi^{(t)}))}{q_{ij}}. \tag{3.38}$$

This can be accomplished by examining all $K!$ permutations for each ν_t $t = 1, ..., M$.

Each of the relabeling algorithms of Stephens shown below corresponds to Algorithm 3.2, for some action a and some loss function L. As Stephens points out, in each of these relabeling algorithms, Step 2 can be solved quickly even for large K.

We will describe two relabeling algorithms of Stephens. In the first, it is desired to cluster the observations into groups. A natural way to do this is to define an $N \times K$ matrix $Q = (q_{ij})$, where q_{ij} represents the probability that observation i is assigned to component j, so each row of Q sums to 1, i.e. every observation is assigned to exactly one component. Following Stephens (1997a, 2000b), we denote the matrix of classification probabilities by $P(\Psi) = (p_{ij}(\Psi))$, where

$$p_{ij}(\Psi) = \frac{w_j f(Y_i|\phi_j)}{\sum_{m=1}^{K} w_m f(Y_i|\phi_m)}, \quad \Psi = (w, \phi). \tag{3.39}$$

Stephens suggested measuring "a loss for reporting Q when the 'true' parameter values are W, Φ" by the Kullback–Leibler distance between the "true" distribution corresponding to $P(\Psi)$ and the distribution corresponding to Q. For this choice of L the general Algorithm 3.2 becomes Algorithm 3.3.

$$L(Q; \Psi) = \sum_{i=1}^{N} \sum_{j=1}^{K} p_{ij}(\Psi) log \left(\frac{p_{ij(\Psi)}}{q_{ij}} \right). \tag{3.40}$$

Stephens (1997a, 2000b) states that it is straightforward to show that

$$\hat{q}_{ij} = \frac{1}{M} \sum_{t=1}^{M} p_{ij}(\nu_t(\Psi^{(t)})). \tag{3.41}$$

There is a neat proof of this result which we give here.

Proof. To minimize D, it is sufficient to minimize

$$D' = -\sum_{t=1}^{M} \sum_{i=1}^{N} \sum_{j=1}^{K} p_{ij}(\nu_t(\Psi^{(t)}))log(q_{ij}).$$

Let $A_{ij} = \sum_{t=1}^{M} p_{ij}(\nu_t(\Psi^{(t)}))$. Then,

$$D' = -\sum_{i=1}^{N} \sum_{j=1}^{K} A_{ij}log(q_{ij}).$$

For each i, $\hat{q}_{ij}$ minimizes the negative of the term in the brackets of the above equation. So fix i and set $x_j = q_{ij}$ and $A_j = A_{ij}$. We need only consider the convex optimization problem:

Minimize $G(x) = -\sum_{j=1}^{K} A_j log(x_j)$ subject to $x > 0$ and $\sum_{j=1}^{K} x_j = 1$.

$G(x)$ is strictly convex as a function of $x = (x_1, ..., x_K)$ on the convex set $x > 0$ and $\sum_{j=1}^{K} x_j = 1$ and hence has a unique global minimum as the solution the the gradient equations. Consider the Lagrange multiplier equation:

$$L(x, \lambda) = -\sum_{j=1}^{K} A_j log(x_j) + \lambda \left(\sum_{j=1}^{K} g_j - 1 \right), \ \lambda > 0.$$

The gradient equations are then

$$\frac{\partial L}{\partial x_j} = -\frac{A_j}{x_j} + \lambda = 0; \ \frac{\partial L}{\partial \lambda} = \sum_{j=1}^{K} x_j - 1 = 0,$$

which implies that

$$\frac{A_j}{\lambda} = x_j.$$

Now summing both sides of this equation we get

$$(1/\lambda) \sum_{j=1}^{K} A_j = \sum_{j=1}^{K} x_j = 1.$$

But

$$\sum_{j=1}^{K} A_j = \sum_{t=1}^{M} \sum_{j=1}^{K} p_{ij}(\nu_t(\Psi^{(t)})) = M$$

so that $\lambda = M$. It follows that

$$\hat{q}_{ij} = x_j = (1/M)A_j = (1/M) \sum_{t=1}^{M} p_{ij}(\nu_t(\Psi^{(t)})),$$

q.e.d.   □

3.5.1   *Linear normal mixture: Eyes problem*

Stephens (1997a) has described another version of Algorithm 3.2 that can
be successfully applied to linear normal mixture models. This version un-
does "the label switching by seeking permutations $\nu_1, ..., \nu_n$ such that the
relabeled sample points agree well on the ordered scale component densities". An outline of the method is given in Algorithm 3.4.

We have applied Algorithm 3.4 to the Eyes problem described by
Eq. (2.35), with initial conditions $\mu = (540, 540)$ using the WinBUGS version of the Gibbs sampler. The first 1,000 iterations were discarded as
burn-in and the last N=10,000 were used for posterior estimation (results
given in Table 3.6). To compare relabeling and re-parametrization strategies, we have computed the Kullback–Leibler distance between the two
kinds of distributions of values μ_1 and μ_2: the distributions obtained as a
result of Stephens' algorithm and the distributions obtained from the Gibbs
sampler of the re-parameterized model (Eq. (3.36)). The Kullback–Leibler
distance will be properly introduced in the next chapter. The Kullback–
Leibler distance is 6×10^{-3} for μ_1 and 9×10^{-4} for μ_2. Thus we can conclude
that Stephens' relabeling algorithm performs well in the framework of our
model.

Depending on the model, Algorithm 3.2 can be computationally intense.
But this is a necessary condition. The common practice is to deal with
label switching by imposing identifiability constraints (i.e. $\mu_1 < ... < \mu_k$ or
$\sigma_1 < ... < \sigma_k$). However, forced ordering of parameters negatively affects
the design and performance of the MCMC [Celeux *et al.* (2000)], thereby
destroying the convergence and proper mixing of the MCMC sampler and
creating bias in the posterior expectation of the parameters.

Algorithm 3.4 Linear normal mixture relabeling algorithm [Stephens (2000b)]

Require: Consider the linear normal mixture model:

$$Y_i \sim \sum_{k=1}^{K} w_k N(\cdot|\mu_k, \Sigma_k). \tag{3.42}$$

Let $\Psi = (w, \mu, \Sigma)$. Assume that $\Psi^{(t)}, t = 1, ..., M$ are sampled values of Ψ from a Markov chain with stationary distribution $p(\Psi|Y)$ Starting with some initial values $\nu_1, ..., \nu_M$, iterate the following steps until a fixed point is reached.

Ensure: 1: Let $\hat{\Psi}$ be given by

$$\hat{w}_i = \frac{1}{M} \sum_{t=1}^{M} w_{\nu_t(i)}^{(t)},$$

$$\hat{\mu}_i = \frac{\sum_{t=1}^{M} w_{\nu_t(i)}^{(t)} \mu_{\nu_t(i)}^{(t)}}{\sum_{t=1}^{M} w_{\nu_t(i)}^{(t)}},$$

$$\hat{\Sigma}_i = \frac{\sum_{t=1}^{M} w_{\nu_t(i)}^{(t)} (\Sigma_{\nu_t(i)}^{(t)} + (\hat{\mu}_i - \mu_{\nu_t(i)}^{(t)})(\hat{\mu}_i - \mu_{\nu_t(i)}^{(t)})^T)}{\sum_{t=1}^{M} w_{\nu_t(i)}^{(t)}},$$

for $i = 1, ..., K$.

Ensure: 2: For $t = 1, ..., M$, choose ν_t to minimize

$$\sum_{i=1}^{K} \{ w_{\nu_t(i)}^{(t)} log|\hat{\Sigma}_i| + w_{\nu_t(i)}^{(t)} trace \left[\hat{\Sigma}_i^{-1} \left(\Sigma_{\nu_t(i)}^{(t)} + (\hat{\mu}_i - \mu_{\nu_t(i)}^{(t)})(\hat{\mu}_i - \mu_{\nu_t(i)}^{(t)})^T \right) \right]$$

$$-2w_{\nu_t(i)}^{(t)} log(\hat{w}_i) - 2(1 - w_{\nu_t(i)}^{(t)}) log(1 - \hat{w}_i) \}.$$

Table 3.6 Eyes example. Output of Stephens' algorithm.

Iteration	w_1	w_2	μ_1	μ_2	σ_1^2	σ_2^2
1	0.502996	0.497	542.433	542.806	49.73	50.7325
2	0.598997	0.401	536.828	548.774	15.7404	16.3435
3	0.598728	0.401272	536.824	548.771	15.7065	16.3647

Chapter 4

Treatment of Mixture Models with an Unknown Number of Components

In this chapter we discuss methods and problems related to the treatment of mixture models with an unknown number of components. The scope of the problem and terminology are outlined in Section 4.1. Next we describe a method of selection of the optimal number of components using weighted Kullback–Leibler distance (Section 4.2) and Stephens' birth–death MCMC (Markov chain Monte Carlo) algorithm (Section 4.3). Finally, we propose a novel method (Section 4.4) for finding the number of mixture components.

4.1   Introduction

Consider a family of mixture probability density functions of the form

$$F^{(K)}(\theta) = \sum_{k=1}^{K} w_k f(\theta|\phi_k), \tag{4.1}$$

where the densities $f(\theta|\phi_k)$ depend on component-specific parameters $\phi_k \in \Phi$, where Φ is a subset of a finite-dimensional Euclidean space. The component weights w_k, $k = 1, ..., K$, also called mixture proportions, have the following properties:

$$w_k \geq 0, \sum_{k=1}^{K} w_k = 1. \tag{4.2}$$

The task is to estimate the unknown number of components K in this mixture. Several powerful methods have been developed for this task: likelihood method [McLachlan *et al.* (1996)], birth–death Markov chain Monte Carlo [Stephens (2000b)], Akaike's information criterion (AIC) [Akaike (1974)], its Bayesian versions, called Bayesian information criterion (BIC) [Schwarz (1978)] and deviance information criterion (DIC) [Spiegelhalter *et al.* (2002)], and Kullback–Leibler distance method [Sahu and Cheng (2003)]. In this book we propose a new method for finding the optimal number of mixture components.

4.2   Finding the Optimal Number of Components Using Weighted Kullback–Leibler Distance

In the framework of the weighted Kullback–Leibler distance method developed by Sahu and Cheng (2003), the optimal number of mixture components is defined as the smallest number of components that adequately explains the structure of the data. This task is traditionally accomplished by the *reductive stepwise method* [Sahu and Cheng (2003)]. In the framework of this method, one starts with the $K = K_0$ component mixture which adequately explains the observed data. Then K is iteratively reduced until the fit is no longer adequate. Reduction of the number of components is achieved by *collapsing* two of the K components (say, $f(\cdot|\phi_{k_1})$ and $f(\cdot|\phi_{k_2})$) by setting $\phi_{k_1} = \phi_{k_2} = \phi^*_{k_1 k_2}$, to be determined.

Adequacy of the fit is assessed using the distance between probability distributions at two consecutive steps. There are several ways to define the distance d between probability density functions. One of them, the Kullback–Leibler distance between two probability densities f and g, is defined as

$$d(f, g) = \int_{S(f)} f(\theta) log \left| \frac{f(\theta)}{g(\theta)} \right| d\theta, \tag{4.3}$$

where $S(f)$ is the support of density f, that is $S(f) = \{\theta : f(\theta) > 0\}$. The Kullback–Leibler distance (also called divergence) was originally introduced by Solomon Kullback and Richard Leibler in 1951 as the directed divergence between two distributions [Kullback and Leibler (1951)]. Unfortunately, in general, the Kullback–Leibler distance cannot be evaluated analytically and Sahu and Cheng suggested using the weighted Kullback–Leibler distance.

Consider two mixture densities, $F^{(K)}(\theta) = \sum_{k=1}^{K} w_k f_k(\theta)$ and $G^{(K)}(\theta) = \sum_{k=1}^{K} w_k g_k(\theta)$, that have the same component weights w_k and $f_k(\theta) = f(\theta|\phi_k)$, $g_k(\theta) = g(\theta|\phi_k)$. The weighted Kullback–Leibler distance $d^*(F^{(K)}, G^{(K)})$ is defined as a weighted sum of the Kullback–Leibler distances $d(f_k, g_k)$ between corresponding components:

$$d^*(F^{(K)}, G^{(K)}) = \sum_{k=1}^{K} w_k d(f_k, g_k). \tag{4.4}$$

As was shown by Sahu and Cheng,

$$d(F^{(K)}, G^{(K)}) \leq d^*(F^{(K)}, G^{(K)}). \qquad (4.5)$$

Now let $F^K(\theta)$ be given by Eq. (4.1). We define $F_{k_1 k_2}^{(K-1)}(\theta)$ to be the *collapsed version* of $F^K(\theta)$ with components k_1 and k_2 merged as follows:

$$F_{k_1 k_2}^{(K-1)}(\theta) = \sum_{k=1}^{K} w_k f(\theta|\phi_k^*), \qquad (4.6)$$

where $\phi_k^* = \phi_k$, $k \neq k_1, k_2$; $\phi_{k_1}^* = \phi_{k_2}^* = \phi_{k_1,k_2}^*$ and ϕ_{k_1,k_2}^* is chosen so that

$$\phi_{k_1,k_2}^* = argmin_\phi\{w_{k_1} d(f(\cdot|\phi_{k_1}), f(\cdot|\phi)) + w_{k_2} d(f(\cdot|\phi_{k_2}), f(\cdot|\phi))\}. \quad (4.7)$$

When $f(\theta|\phi_k) = N(\theta|\mu_k, \Sigma_k)$, $\phi_k = (\mu_k, \Sigma_k)$, an analytical expression can be given for $\phi_{k_1 k_2}^*$, see Section 4.2.2. The same is true for the binomial and Poisson cases, see Section 4.2.6 and Section 4.2.7.

The best collapsed version denoted by $F^{*(K-1)}$ is the one that minimizes the weighted Kullback–Leibler distance $d^*(F^{(K)}, F_{k_1 k_2}^{*(K-1)})$ over all k_1, k_2 such that $k_1 \neq k_2$. Notice that the minimum weighted Kullback–Leibler distance is invariant under permutations of w and ϕ. The K-component model can be replaced by the $(K - 1)$-component model if the distance between the best collapsed version and the original model is less than some cut-off c. Choice of c depends on the data structure and type of distribution used.

4.2.1 *Distance between K-component model and collapsed (K-1)-component model*

From Eq. (4.1) we have:

$$F^{(K)}(\cdot|\phi) = \sum_{k=1}^{K} w_k f(\cdot|\phi_k)$$

$$= \sum_{k \neq \{k_1, k_2\}} w_k f(\cdot|\phi_k) + w_{k_1} f(\cdot|\phi_{k_1}) + w_{k_2} f(\cdot|\phi_{k_2})$$

$$\equiv F_{k \neq \{k_1, k_2\}}^{(K-2)} + F_{k = \{k_1, k_2\}}^{(2)}, \qquad (4.8)$$

$$F_{k_1, k_2}^{*(K-1)}(\cdot|\phi) = \sum_{k \neq \{k_1, k_2\}} w_k f(\cdot|\phi_k) + (w_{k_1} + w_{k_2}) f(\cdot|\phi^*)$$

$$\equiv F_{k \neq \{k_1, k_2\}}^{(K-2)} + F_{k_1, k_2}^{*(1)}.$$

As we can see from the equations above, components $k \neq \{k_1, k_2\}$ are unaffected by the collapse, thus comparison between K- and collapsed $(K-1)$-component models can be reduced to comparison between two-component and one-component models.

Since $F_{k=\{k_1,k_2\}}^{(2)} = w_{k_1} f(\cdot|\phi_{k_1}) + w_{k_2} f(\cdot|\phi_{k_2})$ and $F_{k_1,k_2}^{*(1)} = w_{k_1} f(\cdot|\phi_{k_1,k_2}^*) + w_{k_2} f(\cdot|\phi_{k_1,k_2}^*)$, the distance between K- and collapsed $(K-1)$-component models is

$$d^* \left(F_{k=\{k_1,k_2\}}^{(2)}, F_{k_1,k_2}^{*(1)} \right) = w_{k_1} d(f(\cdot|\phi_{k_1}), f(\cdot|\phi_{k_1,k_2}^*))$$

$$+ w_{k_2} d(f(\cdot|\phi_{k_2}), f(\cdot|\phi_{k_1,k_2}^*)), \qquad (4.9)$$

where

$$\phi_{k_1,k_2}^* = argmin_\phi \{w_{k_1} d(f(\cdot|\phi_{k_1}), f(\cdot|\phi)) + w_{k_2} d(f(\cdot|\phi_{k_2}), f(\cdot|\phi))\}. \quad (4.10)$$

4.2.2 *Weighted Kullback–Leibler distance for the mixture of multivariate normals*

Consider a linear mixture of multivariate normal distributions:

$$F^{(K)}(\cdot|\mu, \Sigma) = \sum_{k=1}^{K} w_k N(\cdot|\mu_k, \Sigma_k), \qquad (4.11)$$

where $\mu = (\mu_1, ..., \mu_K)$ and $\Sigma = (\Sigma_1, ..., \Sigma_K)$. Assuming that we have collapsed components k_1 and k_2. Now

$$F_{k=\{k_1,k_2\}}^{(2)} = w_{k_1} N(\cdot|\mu_{k_1}, \Sigma_{k_1}) + w_{k_2} N(\cdot|\mu_{k_2}, \Sigma_{k_2}),$$

$$\qquad (4.12)$$

$$F_{k_1,k_2}^{*(1)} = (w_{k_1} + w_{k_2}) N(\cdot|\mu^*, \Sigma^*).$$

It is well-known, see Gil *et al.* (2013), that the KL distance between two multivariate normal distributions is given by:

$$d(N(\cdot|\mu_1, \Sigma_1), N(\cdot|\mu_2, \Sigma_2)) = (1/2)\{trace((\Sigma_2^{-1}\Sigma_1)$$

$$+ (\mu_2 - \mu_1)^T (\Sigma_2)^{-1} (\mu_2 - \mu_1) - log(det\Sigma_1/det\Sigma_2) - n\}. \qquad (4.13)$$

Then using Eq. 4.9, the weighted Kullback–Leibler distance between a K-component mixture and its collapsed $(K-1)$-component version is

$$d^*(F_{k=\{k_1,k_2\}}^{(2)}, F_{k_1,k_2}^{*(1)}) = \sum_{\kappa \in \{k_1,k_2\}} \frac{w_\kappa}{2} \{trace((\Sigma^*)^{-1}\Sigma_\kappa)$$

$$+ (\mu^* - \mu_\kappa)^T (\Sigma^*)^{-1} (\mu^* - \mu_\kappa) + log(det(\Sigma^*(\Sigma_\kappa)^{-1})) - n\}.$$

The values of μ^* and Σ^* minimizing $d^* \left(F^{(2)}_{k=\{k_1,k_2\}}, F^{*(1)}_{k_1,k_2} \right)$ are

$$\mu^* = \frac{w_{k_1}\mu_{k_1} + w_{k_2}\mu_{k_2}}{(w_{k_1} + w_{k_2})},$$

(4.14)

$$\Sigma^* = \frac{w_{k_2}\Sigma_{k_2} + w_{k_1}\Sigma_{k_1}}{(w_{k_1} + w_{k_2})} + \frac{w_{k_1}w_{k_2}(\mu_{k_2} - \mu_{k_1})(\mu_{k_2} - \mu_{k_1})^T}{(w_{k_1} + w_{k_2})^2}.$$

Substituting these expressions for μ^* and Σ^* in Eq. (4.14), we get

$$d^* \left(F^{(2)}_{k=\{k_1,k_2\}}, F^{*(1)}_{k_1,k_2} \right)$$

$$= \sum_{\kappa \in \{k_1,k_2\}} \frac{w_\kappa}{2} \{trace((\Sigma^*)^{-1}\Sigma_\kappa) + log(det(\Sigma^*(\Sigma_\kappa)^{-1})) - n\}$$

$$+ \frac{w_{k_1}w_{k_2}}{2(w_{k_1} + w_{k_2})^2}(\mu_{k_1} - \mu_{k_2})^T (\Sigma^*)^{-1}(\mu_{k_1} - \mu_{k_2}).$$

(4.15)

Let $F^{*(K-1)}$ be the collapsed version of $F^{*(K-1)}_{k_1 k_2}$ that minimizes the distance $d(F^{(K)}, F^{*(K-1)}_{k_1 k_2})$ over all (k_1, k_2), $k_1 \neq k_2$. It is important to note that $F^{*(K-1)}$ is a measurable function of w_k, μ_k, Σ_k; $k = 1, ..., K$.

These results are stated in Sahu and Cheng (2003), Lemma 2, without proof. The general proof is given in Appendix C.2. We accept the one-component model if the distance $d^*(F^{(2)}_{k=\{k_1,k_2\}}, F^{*(1)}_{k_1,k_2})$ is small. The choice of the cut-off for collapsing is mainly empirical.

Let us consider a special case, when we erroneously assign too many fixed components. In this case, after fitting the model, we notice that $w_{\tilde{k}} \approx 0$ for unnecessary components of the mixture. Collapsing these components with a real component k, we see that $\mu^* \approx \mu^k$ and $\Sigma^* \approx \Sigma^k$.

$$d^* \left(F^{(2)}_{k=\{\tilde{k},k\}}, F^{*(1)}_{\tilde{k},k} \right) = \sum_{\kappa \in \{k,\tilde{k}\}} \frac{w_\kappa}{2} \{trace((\Sigma^*)^{-1}\Sigma^*) + log(det(\Sigma^*(\Sigma^*)^{-1})) - n\}$$

$$= \sum_{\kappa \in \{k,\tilde{k}\}} \frac{w_\kappa}{2}(n-n) = 0.$$

(4.16)

Therefore, the weighted Kullback–Leibler distance can be used to identify the unnecessary components in a mixture.

4.2.3 *Weighted Kullback–Leibler distance for the multivariate normal mixture model with diagonal covariance matrix*

For a large class of problems, the covariance matrix Σ_k is diagonal. Let $\Sigma_k = (\sigma_k^{(ij)})$, $\sigma_k^{ij} = 0$ if $i \neq j$, and $\sigma_k^{(ii)} \equiv \sigma_k^{(i)}$. Then, $trace((\Sigma^*)^{-1}\Sigma_\kappa) = \sum_{i=1}^{n} \frac{\sigma_\kappa^{(i)\,2}}{\sigma^{(i)*2}}$ and $det(\Sigma^*(\Sigma_\kappa)^{-1}) = \prod_{i=1}^{n} \frac{\sigma^{(i)*2}}{\sigma_\kappa^{(i)2}}$ and $log(\prod_{i=1}^{n}(\frac{\sigma^{*(i)}}{\sigma_\kappa^{(i)}})^2) = \sum_{i=1}^{n} log((\frac{\sigma^{*(i)}}{\sigma_\kappa^{(i)}})^2)$.

The weighted Kullback–Leibler distance between two components will be equal to

$$d\left(F_{k=\{k_1,k_2\}}^{(2)}, F_{k_1,k_2}^{*(1)}\right) = \sum_{\kappa \in \{k_1,k_2\}} \frac{w_\kappa}{2}\left\{\sum_{i=1}^{n}\left[\left(\frac{\sigma_\kappa^{(i)}}{\sigma^{*(i)}}\right)^2 + log\left(\frac{\sigma^{*(i)}}{\sigma_\kappa^{(i)}}\right)^2\right] - n\right\}$$

$$+ \frac{w_{k_1} w_{k_2}}{2(w_{k_1} + w_{k_2})}\sum_{i=1}^{n}\frac{(\mu^{(i)}{}_{k_1} - \mu^{(i)}{}_{k_2})^2}{\sigma^{*2(i)}} \qquad (4.17)$$

$$= \sum_{i=1}^{n} d^{(i)}\left(F_{k=\{k_1,k_2\}}^{(2)}, F_{k_1,k_2}^{*(1)}\right).$$

In this case, the simplified expressions for $\mu^{*(i)}, \sigma^{*(i)^2}$ are

$$\mu^{*(i)} = \frac{w_{k_1}\mu_{k_1}^{(i)} + w_{k_2}\mu_{k_2}^{(i)}}{(w_{k_1} + w_{k_2})},$$

$$\sigma^{*(i)2} = \frac{(w_{k_2}(\sigma_{k_2}{}^{(i)})^2 + w_{k_1}(\sigma_{k_1}{}^{(i)})^2)(w_{k_1} + w_{k_2}) + w_{k_1}w_{k_2}(\mu^{(i)}{}_{k_2} - \mu^{(i)}{}_{k_1})^2}{(w_{k_1} + w_{k_2})^2}.$$

$$(4.18)$$

4.2.4 *Weighted Kullback–Leibler distance for the one-dimensional mixture of normals*

One important special case is a mixture of two one-dimensional normal distributions:

$$F^{(2)}(\cdot) = \sum_{k=1}^{2} w_k N(\cdot|\mu_k, \sigma_k^2). \qquad (4.19)$$

In one dimension, the weighted Kullback–Leibler distance (Eq. (4.15)) is

$$d^* \left(F^{(2)}, F_{12}^{*(1)} \right) = log \left(\frac{\sigma^{*w_1+w_2}}{\sigma_1^{w_1} \sigma_2^{w_2}} \right) + \frac{w_1}{2\sigma^{*2}} (\sigma_1^2 + (\mu^* - \mu_1)^2)$$

$$+ \frac{w_2}{2\sigma^{*2}} (\sigma_2^2 + (\mu^* - \mu_2)^2) - \frac{w_1 + w_2}{2}. \tag{4.20}$$

As a result, expressions for μ^*, σ^* are

$$\mu_{min}^* = \frac{w_1\mu_1 + w_2\mu_2}{w_1 + w_2},$$

$$\sigma_{min}^{*~2} = \frac{w_1\sigma_1^2 + w_2\sigma_2^2}{w_1 + w_2} + \frac{w_1w_2(\mu_1 - \mu_2)^2}{(w_1 + w_2)^2}. \tag{4.21}$$

4.2.5 *Comparison of weighted and un-weighted Kullback–Leibler distances for one-dimensional mixture of normals*

Let us compare the weighted Kullback–Leibler distance to the original un-weighted Kullback–Leibler distance. The original Kullback–Leibler distance for normal mixtures was discussed by Robert (1996), Chapter 24. In his analysis, he used the Kullback–Leibler distance:

$$d(F^{(K)}, G^{(K)}) = \int F^{(K)} log \left| \frac{F^{(K)}}{G^{(K)}} \right| d\theta. \tag{4.22}$$

to compare two- and one-component distributions. These distributions are given by

$$F^{(2)}(\cdot) = w_1 N(\cdot|\mu_1, \sigma_1^2) + w_2 N(\cdot|\mu_2, \sigma_2^2),$$

$$G^{(1)}(\cdot) = N(\cdot|\mu, \sigma^2). \tag{4.23}$$

where $w_1 + w_2 = 1$.

The Kullback–Leibler distance in this case is equal to

$$d(F^{(2)}, G^{(1)}) = \frac{w_1}{\sigma^2}((\mu - \mu_1)^2 + \sigma_1^2) + \frac{w_2}{\sigma^2}((\mu - \mu_2)^2 + \sigma_2^2) + log\sigma^2 + C, \tag{4.24}$$

where $C = \{$terms that do not contain μ or $\sigma\}$.

It is straightforward to show that $d\left(F^{(2)}, G^{(1)} \right)$ is a convex function of $(\mu, 1/\sigma^2)$. Therefore the global minimizer of $d\left(F^{(2)}, G^{(1)} \right)$ is given by the solution (μ, σ^2) of the gradient equations. We have

$$
\frac{\partial d(F^{(2)}, G^{(1)})}{\partial \mu} = \frac{\mu - w_1\mu_1 - w_2\mu_2}{\sigma^2},
$$

$$
\frac{\partial d(F^{(2)}, G^{(1)})}{\partial \sigma^2} = \frac{1}{\sigma^2} - \frac{w_1}{\sigma^4}((\mu - \mu_1)^2 + \sigma_1^2) - \frac{w_2}{\sigma^4}((\mu - \mu_2)^2 + \sigma_2^2).
$$

(4.25)

Setting these derivatives to zero, we get

$$
\mu = w_1\mu_1 + w_2\mu_2,
$$

$$
\sigma^2 = w_1\sigma_1^2 + w_2\sigma_2^2 + w_1w_2(\mu_1 - \mu_2)^2.
$$

(4.26)

This result is the same as obtained when using the weighted Kullback–Leibler distance in Eq. (4.21).

Extension of the original Kullback–Leibler distance to $K > 2$ is more complicated. Thus, the weighted Kullback–Leibler approach developed by Sahu and Cheng is a major simplification.

In the case of the weighted Kullback–Leibler distance, the procedure can be easily extended to mixtures with an arbitrary large number of components K. We treat the comparison between any K- and $(K-1)$-component models with the same simplicity as in the case of a two-component mixture.

Later in the book we will also need the distance formulas for the mixtures of binomial and Poisson distributions. We compute these quantities now.

4.2.6 *Weighted Kullback–Leibler distance for a mixture of binomial distributions*

Consider a mixture of binomial distributions:

$$
F^{(K)}(\cdot) = \sum_{j=1}^{K} w_j f(\cdot|p_j),
$$

(4.27)

where $f(\cdot|p) = Bin(\cdot|n, p)$, $0 < p < 1$ and n is assumed to be fixed.

We now consider a comparison of K-component and $(K-1)$-component mixtures. From Eq. (4.9),

$$
d^*\left(F^{(K)}, F_{k_1, k_2}^{*(K-1)}\right) = \sum_{j=k_1, k_2} w_j d(f(\cdot|p_j), f(\cdot|p^*)).
$$

(4.28)

It is a well-known property of the Kullback–Leibler distance that

$$d(f(\cdot|p), f(\cdot|p^*)) = np\,log\frac{p}{p^*} + n(1-p)log\frac{1-p}{1-p^*}. \tag{4.29}$$

It follows:

$$d^*\left(F^{(K)}, F_{k_1,k_2}^{*(K-1)}\right) = w_{k_1}\left\{np_{k_1}log\frac{p_{k_1}}{p^*} + n(1-p_{k_1})log\frac{1-p_{k_1}}{1-p^*}\right\}$$

$$+ w_{k_2}\left\{np_{k_2}log\frac{p_{k_2}}{p^*} + n(1-p_{k_2})log\frac{1-p_{k_2}}{1-p^*}\right\}. \tag{4.30}$$

It is straightforward to show that $d^*\left(F^{(K)}, F_{k_1,k_2}^{*(K-1)}\right)$ is a convex function of p^*. Therefore the global minimizer of $d^*\left(F^{(K)}, F_{k_1,k_2}^{*(K-1)}\right)$ is given by the solution p^* of the gradient equations. We have

$$\frac{\partial d^*(F^{(K)}, F_{k_1,k_2}^{*(K-1)})}{\partial p^*} = n\{\frac{w_{k_1}p_{k_1} + w_{k_2}p_{k_2}}{p^*(1-p)^*} - \frac{w_{k_1} + w_{k_2}}{1-p^*}\}. \tag{4.31}$$

Setting

$$\frac{\partial d^*(F^{(K)}, F_{k_1,k_2}^{*(K-1)})}{\partial p^*} = 0, \tag{4.32}$$

we get

$$p^* = \frac{w_{k_1}p_{k_1} + w_{k_2}p_{k_2}}{(w_{k_1} + w_{k_2})}. \tag{4.33}$$

4.2.7   *Weighted Kullback–Leibler distance for a mixture of Poisson distributions*

Consider a mixture of Poisson distributions:

$$F_j^{(K)}(\cdot) = \sum_{j=1}^{K} w_j f(\cdot|\lambda), \tag{4.34}$$

where $f(x|\lambda) = \frac{1}{x!}e^{-\lambda}\lambda^x$, for $x = 0, \cdots, \infty$, and $\lambda > 0$,

We now consider a comparison of K-component and $(K-1)$-component mixtures. From Eq. (4.9),

$$d^*\left(F^{(K)}, F_{k_1,k_2}^{*(K-1)}\right) = \sum_{j=k_1,k_2} w_j d(f(\cdot|\lambda_j), f(\cdot|\lambda^*)). \tag{4.35}$$

It is a well-known property of the Kullback–Leibler distance that

$$d(f(\cdot|\lambda), f(\cdot|\lambda^*)) = (\lambda^* - \lambda) - \lambda(log(\lambda^*) - log(\lambda)). \tag{4.36}$$

It follows:

$$d^* \left(F^{(K)}, F_{k_1,k_2}^{*(K-1)} \right) = \sum_{j=k_1,k_2} w_j \left\{ (\lambda^* - \lambda_j) - \lambda_j (log\lambda^* - log\lambda_j) \right\},$$

It is straightforward to show that $d^* \left(F^{(K)}, F_{k_1,k_2}^{*(K-1)} \right)$ is a convex function of λ^* . Therefore the global minimizer of $d^* \left(F^{(K)}, F_{k_1,k_2}^{*(K-1)} \right)$ is given by the solution λ^* of the gradient equations. We have

$$\frac{\partial d^* \left(F^{(K)}, F_{k_1,k_2}^{*(K-1)} \right)}{\partial \lambda^*} = \sum_{j=k_1,k_2} w_j \left\{ 1 - \frac{\lambda_j}{\lambda^*} \right\}.$$

Setting

$$\frac{\partial d^* \left(F^{(K)}, F_{k_1,k_2}^{*(K-1)} \right)}{\partial \lambda^*} = 0, \tag{4.37}$$

we get

$$\lambda^* = \frac{w_{k_1} \lambda_{k_1} + w_{k_2} \lambda_{k_2}}{(w_{k_1} + w_{k_2})}. \tag{4.38}$$

4.2.8   *Metric and semimetric*

In this book we discuss the feasibility of using the weighted Kullback–Leibler distance as a measure of similarity between components of a mixture distribution for the purposes of clustering and classification of objects. Ideally, such a distance measure between objects should be a metric or at least a semimetric [Moen (2000); Oh and Kim (2004)].

A metric is a function that defines a distance between elements of a set. More precisely, a metric on a set X is a function $\tilde{d} : X \times X \to R$, where R is the set of real non-negative numbers. For all elements h, g, z of set X, this function is required to satisfy the following conditions:

(M.1) $\tilde{d}(h,g) = \tilde{d}(g,h)$ (symmetry)
(M.2) $\tilde{d}(h,g) \geq 0$ (non-negativity)
(M.3) $\tilde{d}(h,g) = 0$ if and only if $h = g$
(M.4) $\tilde{d}(h,z) \leq \tilde{d}(h,g) + \tilde{d}(g,z)$ (triangle inequality).

Let f, g and z be probability density functions from the space of the components of some mixture distribution. In our case, the set X consists of all possible pairs (w, f) of probability density functions coupled with their corresponding component weights. Requirements of symmetry and non-negativity follow naturally from the basic concepts of the classification problem. The requirement that $\tilde{d}(h, g) = 0$ if and only if $h = g$ is essential for grouping together identical elements. The triangle inequality is important for efficient treatment of large datasets. For example, if we find that distances $\tilde{d}(g, h)$ and $\tilde{d}(g, z)$ are small, and therefore decide to place elements described by g, z and h in the same cluster, we would like to be sure that $\tilde{d}(h, z)$ is also small. Without the triangle inequality, the distance $\tilde{d}(h, z)$ may be prohibitively large. The triangle inequality would allow us to skip the computation of distances between all pairs of cluster members. Unfortunately, the triangle inequality does not hold for the weighted Kullback–Leibler distance; below we develop an upper bound for $\tilde{d}(h, z)$. Let

$$F^{(K)}(\cdot) = \sum_{k=1}^{K} w_k f(\cdot | \phi_k) \qquad (4.39)$$

be a mixture distribution. Assume the representation for $F^{(K)}$ is not redundant. That is, if $k \neq j$ then $f(\cdot | \phi_k) \neq f(\cdot | \phi_j)$. We let $X = \{(w_k, f(\cdot | \phi_k)) : k = 1, ..., K\}$.

A function that satisfies the first three conditions (M.1–M.3), but not necessarily the triangle inequality (M.4), is called a *semimetric*.

Let us see which of the metric conditions are satisfied in our setting. Let $F^{(K)}$ be given by Eq. (4.39). Define $h = (w_1, f(\cdot | \phi_1))$, $g = (w_2, f(\cdot | \phi_2))$ and $z = (w_3, f(\cdot | \phi_3))$. The kth component is defined by the parameter ϕ_k, and $f_k \equiv f(\cdot | \phi_k)$. The best collapsed $(K - 1)$ version, when components 1 and 2 are collapsed, is defined by the parameter

$$\phi_{1,2}^* = arg \left\{ min_\phi \sum_{k=1,2} w_k E_k log \frac{f(\cdot | \phi_k)}{f(\cdot | \phi)} \right\}; \qquad (4.40)$$

hence $f(\cdot | \phi_{1,2}^*)$ is the corresponding collapsed probability density. Recall that the Kullback–Leibler distance between two functions f_1 and f_2 is defined as $d(f_1, f_2) = \int f_1 log \frac{f_1}{f_2} d\theta$. Then define a distance $\tilde{d}(h, g)$ between two components h and g of the mixture distribution $F^{(K)}$ as

$$\tilde{d}(h, g) = w_1 d(f(\cdot | \phi_1), f(\cdot | \phi_{1,2}^*)) + w_2 d(f(\cdot | \phi_2), f(\cdot | \phi_{1,2}^*)). \qquad (4.41)$$

We have

$$\tilde{d}(h, g) = w_2 d(f(\cdot | \phi_2), f(\cdot | \phi_{1,2}^*)) + w_1 d(f(\cdot | \phi_1), f(\cdot | \phi_{1,2}^*)) = \tilde{d}(g, h). \quad (4.42)$$

Therefore, condition (M.1) is satisfied. To check the conditions (M.2) and (M.3), note that $\tilde{d}(h, g)$ is the weighted sum of two Kullback–Leibler distances. It is well known that

$$d(f, g) \geq 0, \text{ for all } f \text{ and } g \text{ and}$$

$$(4.43)$$

$$d(f, g) = 0 \text{ iff } f = g.$$

Therefore the condition (M.2) is satisfied.

Thus, if $\tilde{d}(h, g) = 0$, then $f(\cdot|\phi_1) = f(\cdot|\phi_2)$. But, by the assumption that $F^{(K)}$ is not redundant, this implies that $w_1 = w_2$ and $h = g$. Therefore the condition (M.3) is satisfied.

It is easy to demonstrate that Kullback–Leibler distance does not satisfy the triangle inequality $d(f_1, f_3) \leq d(f_1, f_2) + d(f_2, f_3)$ by using a counter-example. Let $f_1 = N(0, 1)$, $f_2 = N(0, 2^2)$, and $f_3 = N(0, 3^2)$. Then $d(f_1, f_3) = log(3) - \frac{4}{9} \approx 0.65$, $d(f_1, f_2) = log(2) - \frac{3}{8}$, and $d(f_2, f_3) = log(\frac{3}{2}) - \frac{5}{18}$; $d(h, g) + d(g, z) \approx 0.44 < d(h, z)$, *q.e.d.* The same is true for the weighted Kullback–Leibler distance.

Note that the distance $\tilde{d}(h, z)$ is minimized by choosing the parameters for the best collapsed version:

$$\tilde{d}(h, z) = w_1 d(f_1, f_{1,3}^*) + w_3 d(f_3, f_{1,3}^*) \qquad (4.44)$$

$$= \frac{1}{2}\{w_1 d(f_1, f_{1,3}^*) + w_1 d(f_1, f_{1,3}^*) + w_3 d(f_3, f_{1,3}^*) + w_3 d(f_3, f_{1,3}^*)\}$$

$$\leq \frac{1}{2}\{w_1 d(f_1, f_{1,2}^*) + w_1 d(f_1, f_{2,3}^*) + w_3 d(f_3, f_{1,2}^*) + w_3 d(f_3, f_{2,3}^*)\}.$$

Now let us add two non-negative terms $w_2 d(f_2, f_{1,2}^*)$ and $w_2 d(f_2, f_{2,3}^*)$ to the right-hand side of the inequality:

$$\tilde{d}(h, z) \leq \frac{1}{2}\{w_1 d(f_1, f_{1,2}^*) + w_2 d(f_2, f_{1,2}^*) \qquad (4.45)$$

$$+ w_3 d(f_3, f_{1,2}^*) + w_1 d(f_1, f_{2,3}^*) + w_2 d(f_2, f_{2,3}^*) + w_3 d(f_3, f_{2,3}^*)\}$$

$$= \frac{1}{2}\{\tilde{d}(h, g) + \tilde{d}(h, z) + w_3 d(f_1, f_{2,3}^*) + w_1 d(f_3, f_{1,2}^*)\}.$$

Therefore, in place of the triangle inequality we have

$$\tilde{d}(h, z) \leq \frac{\tilde{d}(h, g) + \tilde{d}(g, z)}{2} + \frac{1}{2}\{w_1 d(f_1, f_{2,3}^*) + w_3 d(f_3, f_{1,2}^*)\}. \qquad (4.46)$$

Therefore, the distance $\tilde{d}$ between the components of a mixture is a semimetric.

4.2.9 Determination of the number of components in the Bayesian framework

Consider now the Bayesian nonlinear mixture model given by Eqs. (2.26)–(2.28):

$$Y_i \sim N(\cdot | h_i(\theta_i), \tau^{-1}\Omega_i),$$

$$\theta_i \sim \sum_{k=1}^{K} w_k N(\cdot | \mu_k, \Sigma_k),$$

$$\mu_k \sim N(\cdot | \lambda_k, \Lambda_k),$$

$$\tau \sim G(\cdot | a, b), \tag{4.47}$$

$$\Sigma_k^{-1} \sim W(\cdot | q_k, \Psi_k),$$

$$w \sim Dir(\cdot | \alpha_1, ..., \alpha_K),$$

and consider n independent vectors $Y_1, ..., Y_n$ generated by this model.

For a fixed number of components K, the MCMC sampler generates random variables $\{\theta^t_i, i = 1, ..., n\}, \tau^t, Q^t = \{w^t_k, \phi^t_k; k = 1, ..., K\}, t = 1, ..., T$. The MCMC sampler is a combination of Gibbs, Metropolis–Hastings, and random permutation sampler, as described in Chapters 2 and 3. Assume that the MCMC sampler approximately converged so that for each t the random vectors $\{\theta^t_i, i = 1, ..., n\}$ are approximately independent and identically distributed from the mixture density:

$$F^{t,(K)}(\theta) = \sum_{k=1}^{K} w^t_k N(\theta | \mu^t_k, \Sigma^t_k). \tag{4.48}$$

For simplicity, from this point on we will suppress the superscript t. To determine the number of components in the mixture $F^{(K)}(\theta) \sim \sum_{k=1}^{K} w_k N(\theta | \mu_k, \Sigma_k)$, we use the formulation of Section 4.2. Select an initial $K = K_0$ and obtain the random variables $d^*(F^{(K)}, F^{*(K-1)})$. We want to estimate the posterior probability:

$$P_{c_K}(K) = P(d^*(F^{(K)}, F^{*(K-1)}) \le c_K | Y_1, ..., Y_n). \tag{4.49}$$

Let

$$\omega^t = I[d^*(F^{(K)}, F^{*(K-1)}) \le c_K], \tag{4.50}$$

where $I[\cdot]$ is the logical indicator function, i.e. $I[True] = 1$, $I[False] = 0$. Then

$$P^*_{c_K} = \frac{1}{T} \sum_{t=1}^{T} \omega^t \tag{4.51}$$

is approximately equal to

$$P_{c_K}(K) = P(d^*(F^{(K)}, F^{*(K-1)}) \leq c_K | Y_1, ..., Y_n). \tag{4.52}$$

If $P^*_{c_K}(K) > \alpha$, then replace $F^{(K)}$ by $F^{*(K-1)}$ and repeat the above process with $K = K - 1$.

If the probability $P^*_{c_K}(K) > \alpha$, the distance between $F^{(K)}$ and $F^{*(K-1)}$ is "small" with "high" probability. The choice of parameters c_K and α is an open question. Sahu and Cheng (2003) suggested using $c_K = \alpha = 0.5$ in the case of a normal mixture. The choice corresponds to the "phase transition" between unimodal and bimodal distributions. If $P^*_{c_K}(K) \leq \alpha$ then stop; further reduction of the number of components is not required.

4.3   Stephens' Approach: Birth–Death Markov Chain Monte Carlo

The birth–death Markov chain Monte Carlo (BDMCMC) method was developed by Matthew Stephens in a series of publications [Stephens (1997a, 2000a)]. In our notation, Stephens considers a linear K-component mixture model of the form

$$Y_i = \theta_i, \tag{4.53}$$

$$\theta_i \sim F^{(K)}(\cdot) = \sum_{k=1}^{K} w_k f(\cdot | \phi_k),$$

with unknown parameters $Q = \{(w_1, \phi_1), ..., (w_K, \phi_K)\}$ and unknown number of components K. In Eq. (4.53), Stephens adds a parameter η which is common to all the components. The posterior distribution of η is then calculated by a standard Gibbs sampler. For simplicity in this presentation we will assume η is known.

Under the BDMCMC methodology, the number of components of the mixture changes dynamically: new components are created (*birth*) or an existing one is deleted (*death*) and model parameters are then recomputed. The Bayesian estimation problem is to calculate the posterior distribution $P(Q, K | Y_1, ..., Y_N)$. Following Stephens (2000a), we assume that birth and death occur as independent Poisson processes with rates $\beta(Q)$ and $\alpha(Q)$, where rates depend on the current state of the process. Probabilities of birth and death are

$$P_b = \frac{\beta(Q)}{\beta(Q) + \alpha(Q)}, \ P_d = \frac{\delta(Q)}{\beta(Q) + \alpha(Q)} = 1 - P_b. \tag{4.54}$$

If the birth of a component with (w, ϕ) occurs, then the process "jumps" from the K-component state $Q = \{(w_1, \phi_1), ..., (w_K, \phi_K)\}$ to the $(K+1)-$ component state

$$Q \cup (w, \phi) = \{(w_1(1-w), \phi_1), ..., (w_K(1-w), \phi_K), (w, \phi)\}. \qquad (4.55)$$

If the death of the k_0^{th} component occurs, then the process "jumps" from the K-component state $Q = \{(w_1, \phi_1), ..., (w_K, \phi_K)\}$ to the $(K-1)-$ component state

$$Q \setminus (w_{k_0}, \phi_{k_0})$$

$$= \left\{ (\frac{w_1}{1-w_{k_0}}, \phi_1), \cdots, (\frac{w_{k_0-1}}{1-w_{k_0}}, \phi_{k_0-1}), (\frac{w_{k_0+1}}{1-w_{k_0}}, \phi_{k_0+1}), \cdots, (\frac{w_K}{1-w_{k_0}}, \phi_K) \right\}.$$
$$(4.56)$$

Note that birth and death moves do not violate the constraint $\sum_{k=1}^{K} w_k = 1$.

As in previous works by Zhou (2004) and Stephens (2000b), we assume a truncated Poisson prior on the number of components K with approximate mean parameter λ:

$$P(K) \propto \frac{\lambda^K e^{-\lambda}}{K!}, \quad K = 1, ..., K_{max}. \qquad (4.57)$$

If we fix the birth rate $\beta(Q) = \lambda_b$, the death rate for each component $k \in \{1, ..., K\}$ is calculated as

$$\alpha_k(Q) = \lambda_b \frac{L(Q \setminus (w_k, \phi_k))}{L(Q)} \frac{P(K-1)}{K P(K)} = \frac{L(Q \setminus (w_k, \phi_k))}{L(Q)} \frac{\lambda_b}{\lambda}, \qquad (4.58)$$

where $L(Q)$ is the likelihood of N independent observations given the model parameters:

$$L(Q) = P(Y_1, ..., Y_N | Q, K) = \prod_{j=1}^{N} \sum_{k=1}^{K} w_k f(Y_j | \phi_k). \qquad (4.59)$$

To construct the BDMCMC sampler we have used Algorithms 3.1 and 3.2 from Stephens (2000a) and the Spiegelhalter *et al.* (2003) implementation of the Gibbs sampler. The summary of our implementation of this approach is given in Algorithm 4.1.

Algorithm 4.1 BDMCMC

Require: Set the number of components $K = K_{init}$, the initial state $Q = \{(w_1, \phi_1), ..., (w_K, \phi_K)\}$, and the fixed birth rate $\beta(Q) = \lambda_b$. Assume a truncated Poisson prior for the number of components K: $P(K) \propto \frac{\lambda^K}{K!} e^{-\lambda}$, for $1 \leq K \leq K_{max}$, and $\lambda > 0$.

Ensure: 1: Run the Gibbs sampler for M iterations.

Ensure: 2: Based on the Gibbs sampler output, calculate parameters for the trans-dimensional step. The death rate $\alpha_k(Q)$ for each component $k = 1, ..., K$ is

$$\alpha_k(Q) = \frac{L(Q \setminus (w_k, \phi_k))}{L(Q)} \frac{\lambda_b}{\lambda}. \tag{4.60}$$

Calculate the total death rate $\alpha(Q) = \sum_{k=1}^{K} \alpha_k(Q)$.

Simulate the time to the next jump from an exponential distribution with mean

$$\frac{1}{\lambda_b + \alpha_k(Q)}. \tag{4.61}$$

Ensure: 3: Simulate the type of jump: birth or death with probabilities $P(birth) = P_b$ and $P(death) = P_d$ where

$$P_b = \frac{\beta(Q)}{\beta(Q) + \alpha(Q)}, \quad P_d = \frac{\alpha(Q)}{\beta(Q) + \alpha(Q)}. \tag{4.62}$$

Ensure: 4: Adjust prior distributions for the Gibbs sampler to reflect birth or death.

Ensure: 5: Repeat **Steps 1–4**.

To simulate the birth or death in Step 4 above, use the following procedure: If $1 < K < K_{max}$, simulate the birth–death process with $P(birth) = P_b$ and $P(death) = P_d$. Generate random number $u \sim U[0, 1]$.

(1) If $P_b > u$, simulate birth of a component (w, ϕ) by simulating new component weight $w \sim Beta(1, K)$ and $\phi \sim p(\phi)$.

(2) If $P_b \leq u$, choose a component to die with probability $\frac{\alpha_k(Q)}{\alpha(Q)}$. This is achieved by computing the cumulative death probability $D(k)$, $k = 1, ..., K$, equal to the probability to kill a component with an index $\leq k$, and generation of a random number $v \sim U[0, 1]$. The component k_0 is chosen to die if $D(k_0 - 1) < v \leq D(k_0)$. Weights of the remaining $K - 1$ components are adjusted as $\frac{w_1}{(1 - w_K)}, ..., \frac{w_{K-1}}{(1 - w_K)}$.

If $K = 1$, simulate birth of a component (w, ϕ).

If $K = K_{max}$, choose a component to die.

It is shown in Stephens (2000a) that under suitable hypotheses, the BDMCMC Algorithm 4.1 converges to the posterior distribution $P(Q, K|Y_1, ..., Y_N)$.

It is relatively straightforward to implement the BDMCMC algorithm using WinBUGS. To make the trans-dimensional steps, we have to stop the WinBUGS operation, save that state, perform the birth/death process, and then re-start WinBUGS with the new state. This is accomplished by integration of WinBUGS into a C++ or R code using the *BackBUGS* script of WinBUGS that allows running BUGS simulations "in the background". In this mode, the CODA output of the simulated parameters is redirected to the user-specified files. The C++ or R program reads the CODA output files, computes birth/death rates and initial values for the next iteration of WinBUGS, and starts another round of the Gibbs sampler.

4.3.1 *Treatment of the Eyes model*

We have applied the BDMCMC algorithm to the Eyes model:

$$
\begin{cases}
Y_i = \theta_i, \\
\theta_i \sim \sum_{k=1}^{K} w_k N(\cdot|\mu_k, \sigma^2), \text{ for } i = 1, ..., N, \\
\mu_k \sim N(\cdot|\mu_0, \sigma_0^2), \text{ for } k = 1, ..., K, \\
w \sim Dir(\cdot|\alpha), \\
K \sim TruncatedPoisson(\cdot|\lambda, K_{max}), \\
\tau = 1/\sigma^2 \sim G(\cdot|a, b), \\
K_{max} = 10, \lambda = 3.
\end{cases}
\tag{4.63}
$$

We start the Gibbs sampler by setting $\mu_0 = 540$, $\sigma_0^2 = 10^3$, $\sigma^2 = 10^3$, $a = 10$, and $b = 100$. The likelihood L is given by the formula

$$
L(Q) = \prod_{i=1}^{N} \sum_{k=1}^{K} w_k N(Y_i|\mu_k, \sigma^2).
\tag{4.64}
$$

The death rate is thus equal to

$$
\alpha_{k_0} = \frac{L(Q \setminus (w_{k_0}, \phi_{k_0}))}{L(Q)} \frac{\lambda_b}{\lambda}
$$

$$
= \frac{\lambda_b}{\lambda} \prod_{i=1}^{N} \frac{\sum_{k=1, k \neq k_0}^{K} w_k N(Y_i | \mu_k, \sigma^2)}{\sum_{k=1}^{K} w_k N(Y_i | \mu_k, \sigma^2)} \tag{4.65}
$$

$$
= \frac{\lambda_b}{\lambda} \prod_{i=1}^{N} \left\{ 1 - \frac{w_{k_0} e^{-\frac{(Y_i - \mu_{k_0})^2}{2\sigma^2}}}{\sum_{k=1}^{K} w_k e^{-\frac{(Y_i - \mu_k)^2}{2\sigma^2}}} \right\}
$$

Following Stephens (2000b), we use $\lambda = \lambda_B$, thus

$$
\alpha_{k_0} = \prod_{i=1}^{N} \left\{ 1 - \frac{w_{k_0} e^{-\frac{(Y_i - \mu_{k_0})^2}{2\sigma^2}}}{\sum_{k=1}^{K} w_k e^{-\frac{(Y_i - \mu_k)^2}{2\sigma^2}}} \right\}
$$

$$
\alpha(Q) = \sum_{k_0}^{K} \alpha_{k_0} \tag{4.66}
$$

$$
P_b = \frac{\lambda_B}{\lambda_B + \alpha(Q)}, \quad P_d = \frac{\alpha(Q)}{\lambda_B + \alpha(Q)},
$$

and the time to the next jump has an exponential distribution with mean $\frac{1}{\lambda + \alpha(Q)}$. For the Eyes dataset of 48 observations, the BDMCMC algorithm needs the following modifications:

(1) Simulation of the component mean is done using $\mu \sim N(\cdot | \mu_0, \sigma_0^2)$.
(2) Set *M=500* iterations.
(3) Use simplified expression for the death rate in Eq. (4.66).
(4) At Step 4 of BDMCMC Algorithm, adjust prior distributions for the Gibbs sampler in the following fashion: for each component, set μ_k to equal the posterior estimate of $\overline{\lambda}_k$; adjust parameters α of the Dirichlet distribution: $\alpha_{k_1} : \alpha_{k_2} = \overline{\pi}_{k_1} : \overline{\pi}_{k_2}$, set $a = \overline{\tau}^2 * 1000$ and $b = \overline{\tau} * 1000$, so that $E(\tau) = a/b$ and $Var(\tau) = a/b^2 = 0.001$.

We ran the above BDMCMC algorithm for 10,000 iterations. The Markov chain was in the two-component state 66% of the time. For $K = 2$ the posterior distributions of parameters are shown in the Figure 4.1. Note, that the Markov chain did not converge, since the distribution of parameters μ and w for the two components differ slightly.

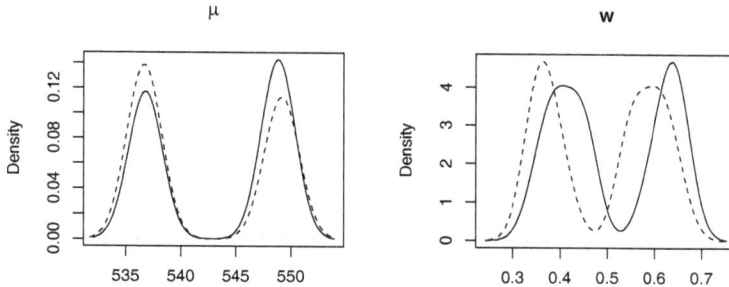

Fig. 4.1.   Birth–death Eyes Model. Distribution of component means and weights for the two-component model, solid and dashed lines correspond to different components.

4.3.2   *Treatment of nonlinear normal mixture models*

Our main interest in this book is studying nonlinear mixture models. The main problem that these models have in applying BDMCMC is in the calculation of the likelihood $L(Q)$. Consider again the general nonlinear mixture model given in Eq. (4.47). If we let $\phi_k = (\mu_k, \Sigma_k)$ and as before let $Q = \{(w_1, \phi_1), ..., (w_K, \phi_K)\}$ then the likelihood $L(Q)$ is given by

$$L(Q) = P(Y_1, ..., Y_N | Q, K),$$

$$= \prod_{i=1}^{N} \sum_{k=1}^{K} w_k \left\{ \int N(Y_i | h_i(\theta_i), \tau^{-1}\Omega_i) N(\theta_i | \mu_k, \Sigma_k) d\theta_i \right\}. \tag{4.67}$$

Consequently the integral $\int N(Y_i | h_i(\theta_i), \tau^{-1}\Omega_i) N(\theta_i | \mu_k, \Sigma_k) d\theta_i$ has to be evaluated at Steps 2–4 of BDMCMC. It is customary to perform these steps thousands of times. Using a numerical procedure such as importance sampling, each integration requires thousands of evaluations [Wang *et al.* (2007, 2009)]. This is a serious handicap of the birth–death method and any method requiring likelihood evaluation. We develop a method that avoids this problem in the next section.

4.4   Kullback–Leibler Markov Chain Monte Carlo – A New Algorithm for Finite Mixture Analysis

The new trans-dimensional method is motivated by several factors. The BDMCMC method calls for repeated evaluation of the likelihood. When the number of observations is large, the likelihood can become very small, resulting in singularities in the death rate. For nonlinear models, evaluation of the likelihood involves integration, which can negatively affect the performance of the algorithm.

One of the most important properties of the Gibbs sampler is that calculating the full conditionals does not require calculating the likelihood. This property is essential for nonlinear pharmacokinetic models where calculation of the likelihood requires high-dimensional integration over the model parameters. Since this integration is required at essentially every iteration of the Gibbs sampler, the resulting algorithm would be unacceptably slow. On the other hand, for linear models the likelihood can be evaluated without integration and the problem disappears. An excellent discussion of this issue can be found in Lunn *et al.* (2002).

The work of Sahu and Cheng (2003) led us to combine the method of choosing the optimal number of components using the weighted Kullback–Leibler distance, the BDMCMC approach outlined in Stephens (2000b), and the random permutation sampler (RPS) in Fruhwirth-Schnatter (2001). Computation of the weighted Kullback–Leibler distance is straightforward for some popular families of distributions (i.e. normal and gamma); its complexity does not depend on the number of observations; nonlinear models can be handled as efficiently as linear ones.

The weighted Kullback–Leibler distance is a natural measure of closeness between $(K+1)$- and K-component models (see Sahu and Cheng (2003) for discussion). The probability distribution of the weighted Kullback–Leibler distance between $(K + 1)$- and K-component models characterizes the current state of the Gibbs sampler. In Bayesian statistics the weighted Kullback–Leibler distance is sometimes used as a measure of the information gain in moving from a prior distribution to a posterior distribution [Dowe *et al.* (1998); Fitzgibbon *et al.* (2002)].

We replace the likelihood calculation required in BDMCMC by a measure based on the weighted Kullback–Leibler distance. Using the notation of Section 4.2.8, we compute the distance $\tilde{d}(k, k')$ between all pairs of components of the mixture distribution $F^{(K)}$, and record the smallest of all distances, $\tilde{d}^*(k) = min\{\tilde{d}(k, k') : k' = 1, ..., K; k' \neq k\}$.

A large value of $\tilde{d}^*(k)$ indicates that the kth component differs from all other components, and the kth component is not a good candidate to be removed. Therefore, the probability to "kill" this component must be small. Conversely, when $\tilde{d}^*(k)$ is small, it means that the kth component does not differ from all other components, and it is a good candidate to be removed. Hence, we need to make the death rate for the kth component of the mixture large.

For this reason, when the Gibbs sampler is at state Q, we define the death rate $\alpha_k(Q)$ for the kth component of $F^{(K)}$ to be the inverse of $\tilde{d}^*(k)$, i.e. $\alpha_k(Q) = 1/\tilde{d}^*(k)$, and the total death rate $\alpha(Q)$ to be equal to $\alpha(Q) = \sum_{k=1}^{K} \frac{1}{\tilde{d}^*(k)}$. Hence, the probability of the death move is $P_d = \frac{\alpha(Q)}{\alpha(Q)+\lambda_B}$, where λ_B is assumed to be equal to the expected number of components, the probability of the birth move is $P_d = \frac{\lambda_B}{\alpha(Q)+\lambda_B}$. The summary of the KLMCMC approach is shown in Algorithm 4.2.

Algorithm 4.2 KLMCMC

Require: Set the number of components $K = K_{init}$, the initial state $Q = \{(w_1, \phi_1), ..., (w_K, \phi_K)\}$, the fixed birth rate $\beta(Q) = \lambda_B$. Assume a truncated Poisson prior for the number of components K: $P(K) \propto \frac{\lambda^K}{K!}e^{-\lambda}$, for $1 \leq K \leq K_{max}$, and $\lambda > 0$.

Ensure: 1: Run the Gibbs sampler for M iterations.

Ensure: 2: Based on the Gibbs sampler output, calculate parameters for the trans-dimensional step. The death rate $\alpha_k(Q)$ for each component $k = 1, ..., K$ is

$$\alpha_k(Q) = \frac{1}{\tilde{d}^*(k)}. \tag{4.68}$$

Calculate the total death rate $\alpha(Q) = \sum_{k=1}^{K} \alpha_k(Q)$.

Ensure: 3: Simulate the type of jump: birth or death with probabilities $P(birth) = P_b$ and $P(death) = P_d$ where

$$P_b = \frac{\beta(Q)}{\beta(Q) + \alpha(Q)}, \ P_d = \frac{\alpha(Q)}{\beta(Q) + \alpha(Q)}. \tag{4.69}$$

Ensure: 4: Adjust prior distributions for the Gibbs sampler.

Ensure: 5: Repeat **Steps 1–4**.

To simulate the birth or death in Step 4 above, use the following algorithm: If $1 < K < K_{max}$, simulate the birth–death process with $P(birth) = P_b$ and $P(death) = P_d$. Generate random number $u \sim U[0, 1]$.

(1) If $P_b > u$, simulate birth of a component (w, ϕ) by simulating new component weight $w \sim Beta(1, K)$ and $\phi \sim p(\phi)$.

(2) If $P_b \leq u$, choose a component to die with probability $\frac{\alpha_k(Q)}{\alpha(Q)}$. This is achieved by computing the cumulative death probability $D(k)$, $k = 1, ..., K$, equal to the probability of killing a component with an index $\leq k$, and generation of a random number $v \sim U[0, 1]$. The component k_0 is chosen to die if $D(k_0 - 1) < v \leq D(k_0)$. Weights of the remaining $K - 1$ components are adjusted as $\frac{w_1}{(1-w_K)}, ..., \frac{w_{K-1}}{(1-w_K)}$.

If $K = 1$, simulate birth of a component (w, ϕ).

If $K = K_{max}$, choose a component to die.

We will refer to this algorithm as *KLMCMC*.

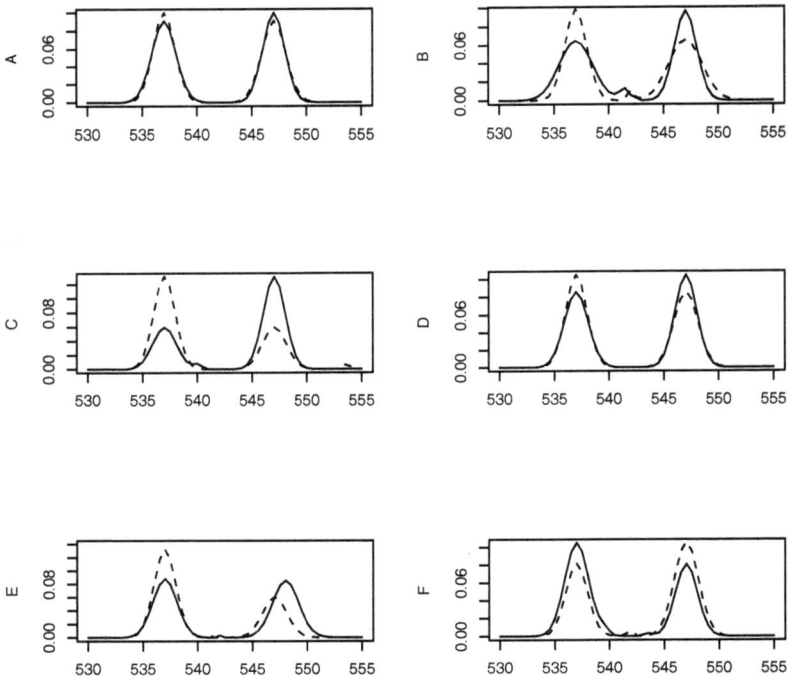

Fig. 4.2.   Eyes example, KLMCMC algorithm. Posterior density of component means μ for $K = 2$ for various parameter values: (A) $\tau_0 = 0.01$, $\lambda_B = 3$; (B) $\tau_0 = 0.01$, $\lambda_B = 2$; (C) $\tau_0 = 0.001$, $\lambda_B = 3$; (D) $\tau_0 = 0.027$, $\lambda_B = 3$, $\sigma = 3.5$ fixed; (E) $\tau_0 = 0.027$, $\lambda_B = 3$, $\sigma = 5$ fixed; (F) $\tau_0 = 0.027$, $\lambda_B = 3$.

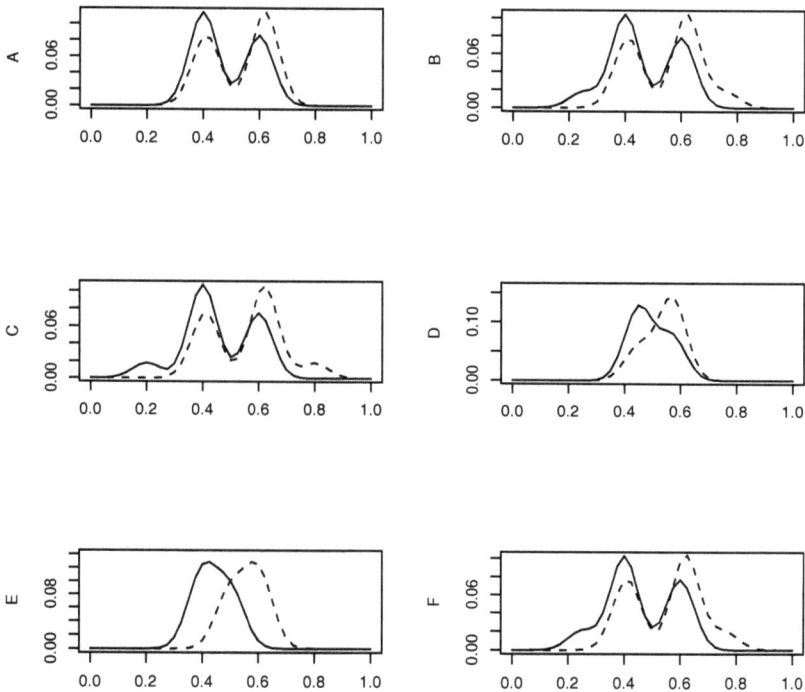

Fig. 4.3.   Eyes example, KLMCMC algorithm. Posterior density of component weights w for $K = 2$ for various parameter values: (A) $\tau_0 = 0.01$, $\lambda_B = 3$; (B) $\tau_0 = 0.01$, $\lambda_B = 2$; (C) $\tau_0 = 0.001$, $\lambda_B = 3$; (D) $\tau_0 = 0.027$, $\lambda_B = 3$, $\sigma = 3.5$ fixed; (E) $\tau_0 = 0.027$, $\lambda_B = 3$, $\sigma = 5$ fixed; (F) $\tau_0 = 0.027$, $\lambda_B = 3$.

Table 4.1   Eyes example. Results of Stephens' relabeling algorithm 3.4 applied to the KLMCMC output.

Parameters	w_1	w_2	μ_1	μ_2	σ
$\tau_0 = 0.01$, $\lambda_B = 3$	0.41	0.59	548.8	536.8	3.60
$\tau_0 = 0.01$, $\lambda_B = 2$	0.38	0.62	547.4	538.1	3.89
$\tau_0 = 0.001$, $\lambda_B = 3$	0.42	0.58	547.5	537.4	3.59
$\tau_0 = 0.027$, $\lambda_B = 3$, $\sigma = 3.5$	0.43	0.57	548.7	536.6	3.50
$\tau_0 = 0.027$, $\lambda_B = 3$, $\sigma = 5$	0.42	0.58	547.5	537.3	5.0
$\tau_0 = 0.027$, $\lambda_B = 3$	0.37	0.59	547.5	537.9	3.98

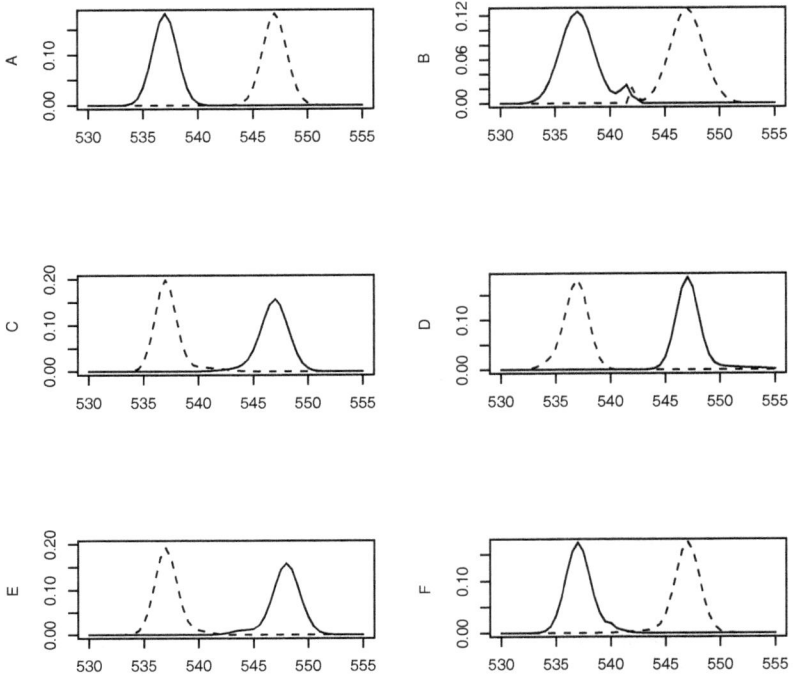

Fig. 4.4. Eyes example, KLMCMC algorithm. Posterior density of component means μ for $K = 2$ after relabeling for various parameter values: (A) $\tau_0 = 0.01$, $\lambda_B = 3$; (B) $\tau_0 = 0.01$, $\lambda_B = 2$; (C) $\tau_0 = 0.001$, $\lambda_B = 3$; (D)$\tau_0 = 0.027$, $\lambda_B = 3$, $\sigma = 3.5$ fixed; (E) $\tau_0 = 0.027$, $\lambda_B = 3$, $\sigma = 5$ fixed; (F) $\tau_0 = 0.027$, $\lambda_B = 3$.

We applied the KLMCMC algorithm to a special case of linear normal mixtures, the Eyes model. For such models, the birth of a new component with parameters (w, μ) is simulated by

$$w \sim Beta(\cdot|1, K), \quad \mu \sim N(\cdot|\mu_0, \sigma_0^2), \tag{4.70}$$

where the values of μ_0 and σ_0^2 are estimated as the mean and variance of the distribution of observations. An easy way to simulate a w from the $Beta(\cdot|1, k)$ distribution is by simulating $Y_1 \sim G(\cdot|1, 1)$ and $Y_2 \sim G(\cdot|k, 1)$ and setting $w = \frac{Y_1}{Y_1+Y_2}$.

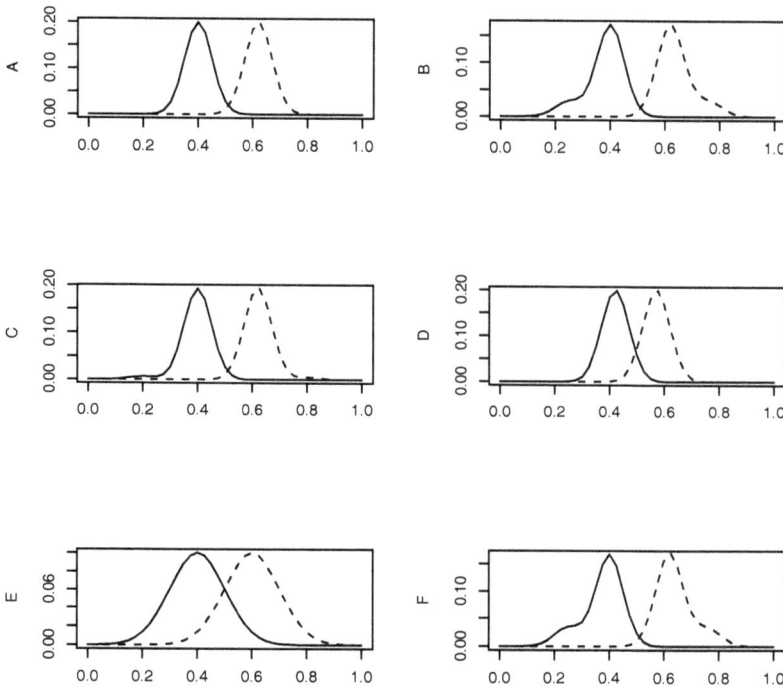

Fig. 4.5.   Eyes example, KLMCMC algorithm. Posterior density of component weights w for $K = 2$ after relabeling for various parameter values: (A) $\tau_0 = 0.01$, $\lambda_B = 3$; (B) $\tau_0 = 0.01$, $\lambda_B = 2$; (C) $\tau_0 = 0.001$, $\lambda_B = 3$; (D)$\tau_0 = 0.027$, $\lambda_B = 3$, $\sigma = 3.5$ fixed; (E) $\tau_0 = 0.027$, $\lambda_B = 3$, $\sigma = 5$ fixed; (F) $\tau_0 = 0.027$, $\lambda_B = 3$.

We tested the method with a wide range of parameters. The expected number of components ranges from $\lambda = \lambda_B = 2$ to $\lambda = \lambda_B = 5$; precision of the prior distribution of the component means τ_0 ranges from 0.027 to 0.001. We also tested various times between the birth–death steps (number of Gibbs sampler iterations in WinBUGS), from $M = 100$ to $M = 2000$. In all cases, approximately 50% of the time the Markov chain was in the state $K = 2$. We followed the run of KLMCMC by the relabeling process (Algorithm 3.2, Section 3.5). The results of KLMCMC and relabeling are shown in Figures 4.2, 4.3, 4.4, and 4.5. The convergence of the KLMCMC process depends on the selection of parameter values. However, the relabeling procedure successfully dealt with all considered datasets (see Table 4.1).

Table 4.2   Eyes example. Output of the KLMCMC with random permutation sampler.

Iterations	w_1	w_2	μ_1	μ_2
1	0.44	0.56	544.81	538.82
2	0.39	0.61	548.72	536.75

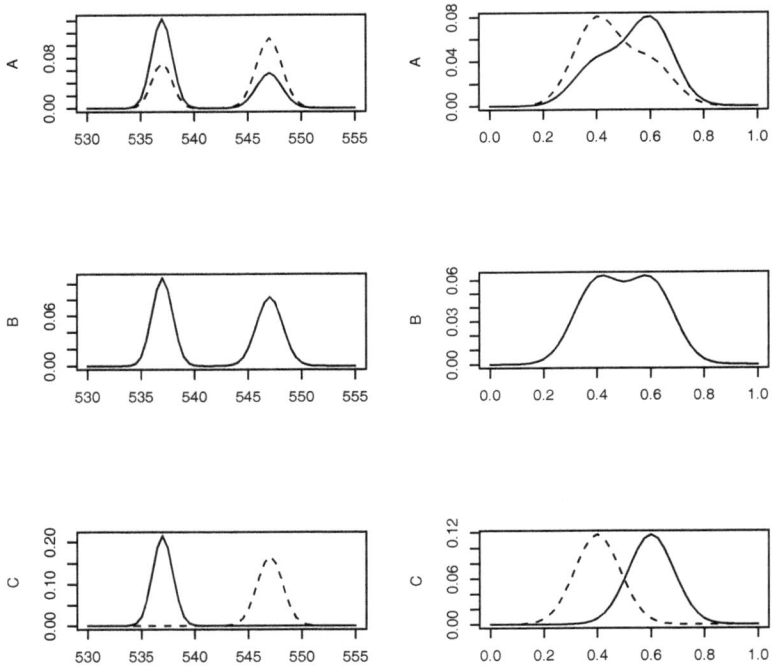

Fig. 4.6.   Eyes example, KLMCMC algorithm. Posterior densities of μ (left) and w (right) for $K = 2$ after relabeling for parameter values $\tau_0 = 0.001$, $\lambda_B = 2$, and fixed $\sigma = 3.5$: (A) KLMCMC output; (B) after random permutations; (C) after relabeling.

4.4.1   *Kullback–Leibler Markov chain Monte Carlo with random permutation sampler*

The mixing and convergence of a Markov chain can be significantly improved by using the random permutation approach suggested by Fruhwirth-Schnatter (2001). At each birth–death step, components of the mixture are randomly permuted (using the random permutation sampler described in Section 3.3), producing a symmetric posterior distribution of estimated model parameters. Stephens' relabeling algorithm described in Section 3.5, Algorithm (3.2), is used to infer model parameters for individual components.

Comparing these results (Table 4.2 and Figure 4.6) with the output of the BDMCMC followed by the relabeling step, we can conclude that the KLMCMC performs as well as the BDMCMC for the normal linear mixture model with the Eyes dataset. The random permutation step improves the performance of the KLMCMC. However, the advantages of the KLMCMC approach become more pronounced when we apply it to nonlinear datasets.

Chapter 5

Applications of BDMCMC, KLMCMC, and RPS

In this chapter, we present several examples of how birth–death MCMC (BDMCMC), Kullback–Leibler MCMC (KLMCMC), and random permutation sampler (RPS) deal with the problem of the selection of the optimal number of mixture components. As examples we consider: Galaxy dataset (Section 5.1), nonlinear simulated mixture dataset (Section 5.2), two-component Boys and Girls dental measurements dataset (Section 5.3), one-compartment pharmacokinetics model (Section 5.4), and seed development dataset (Section 5.5). In referring to the algorithms used we mean specifically: BDMCMC – Algorithm 4.1, KLMCMC – Algorithm 4.2, RPS – Algorithm 3.1, Stephens' relabeling – Algorithm 3.2 for linear normal mixture models, otherwise Algorithm 3.3. Our goal in each case is to determine the number of mixture components and to estimate the unknown values of component-specific parameters.

5.1 Galaxy Data

In this section we analyze the famous Galaxy dataset initially presented by Postman *et al.* (1986). This dataset contains recession velocities of 82 galaxies from six separated conic sections of space, ordered from smallest to largest and scaled by a factor of 1,000. This dataset has been analyzed by several authors, including Roeder (1990), Escobar and West (1995), Carlin and Chib (1995), Phillips and Smith (1996), Stephens (1997a,b, 2000a,b), Richardson and Green (1997), Aitkin (2001) and Cappé *et al.* (2003).

The popularity of this dataset is attributed to an interesting problem in modern astrophysics: analysis of gravitational clusters in the universe. It is known that the distribution of galaxies in space is strongly clustered. Astrophysicists have developed general models for the formation of

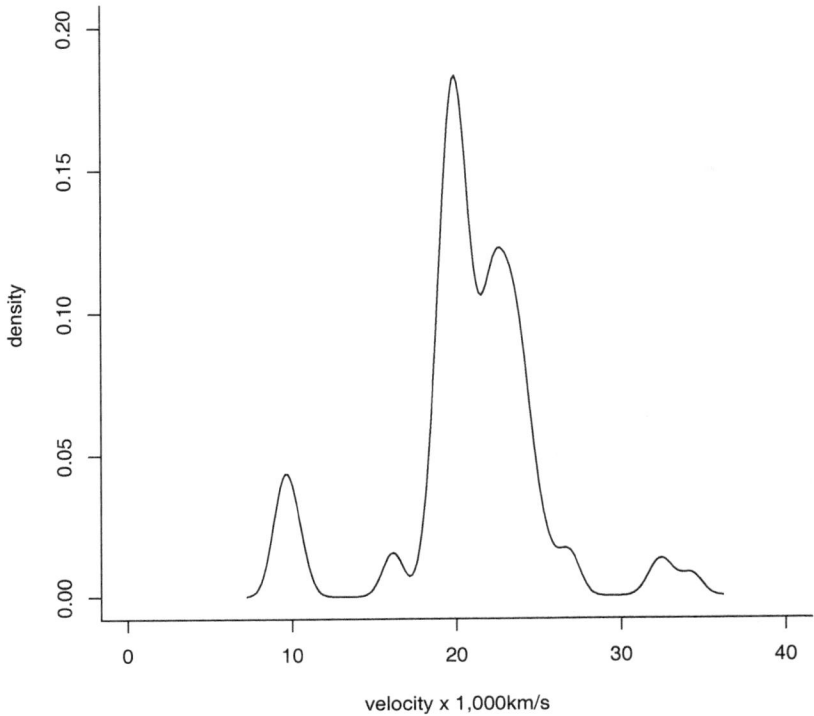

Fig. 5.1.　Galaxy data.　Distribution of velocities (measured in 10^3 km/s) of distant galaxies diverging from our galaxy, from six well-separated conic sections of the Corona Borealis.

gravitational clusters. However, the details of galaxy clustering, processes of collision and merging clusters are not yet well studied. Optical measurements of galaxy red shifts (recessional velocities) can be used to determine the distance from a galaxy to Earth. Since the Big Bang, the universe has been expanding, and the recessional velocity follows Hubble's Law $v = H_0 d$, where v is the velocity in km/s, d is the galaxy distance from the Earth in megaparsecs (Mpc), and $H_0 \approx 72$ km/s/Mpc is the Hubble constant [Drinkwater *et al.* (2004)]. Therefore, the number of clusters in the recessional velocities of galaxies can be an indication of the number of clusters of galaxies in space [Roeder (1990); Aitkin (2001)].

　　The Galaxy dataset listed in Table 5.1 and graphed in Figure 5.1 contains velocities of 82 distant galaxies, measured in 10^3 km/s, diverging from

Table 5.1   Galaxy dataset [Postman *et al.* (1986)], containing recession velocities of 82 galaxies.

9.172	9.350	9.483	9.558	9.775	10.227	10.406	16.084
16.170	18.419	18.552	18.600	18.927	19.052	19.070	19.330
19.343	19.343	19.440	19.473	19.529	19.541	19.547	19.663
19.846	19.856	19.863	19.914	19.918	19.973	19.989	20.166
20.175	20.179	20.196	20.215	20.221	20.415	20.629	20.795
20.821	20.846	20.875	20.986	21.137	21.492	21.701	21.814
21.921	21.960	22.185	22.209	22.242	22.249	22.314	22.374
22.495	22.746	22.747	22.888	22.914	23.206	23.241	23.263
23.484	23.538	23.542	23.666	23.706	23.711	24.129	24.285
24.289	24.366	24.717	24.990	25.633	26.960	26.995	32.065
32.789	34.279						

our galaxy from six well-separated conic sections of the Corona Borealis. The relative measurement error is estimated to be less than 0.5%, or 0.05 units [Escobar and West (1995); Aitkin (2001)].

In 1990, Roeder (1990) used the Galaxy dataset to support her theoretical developments. She assumed that each velocity of a galactic cluster can be described by a normal distribution. In our notation the Galaxy model is given by

$$Y_i \sim \sum_{k=1}^{K} w_k N(\cdot|\mu_k, \sigma_k^2),$$

$$\mu_k \sim N(\cdot|\mu_0, \sigma_0^2),$$

$$\sigma_k^{-2} \sim G(\cdot|\vartheta, \beta), \qquad (5.1)$$

$$\beta \sim G(\cdot|\delta, \epsilon),$$

$$w \sim Dir(\cdot|\alpha).$$

The hyper-parameters are $\{\mu_0, \sigma_0, \vartheta, \delta, \epsilon, \alpha\}$ and they are defined below. Visual inspection of the dataset (Figure 5.1) suggests that there are at least $K = 3$ components. As was discussed by Aitkin (2001), different reasonable prior modeling approaches result in different conclusions about the number of components: the smallest reported value was three and the largest was nine components. Apparently, prior structures have profound effects on model likelihoods and posterior parameter estimates.

Our goal is to estimate the number of components and the values of the unknown parameters in the model.

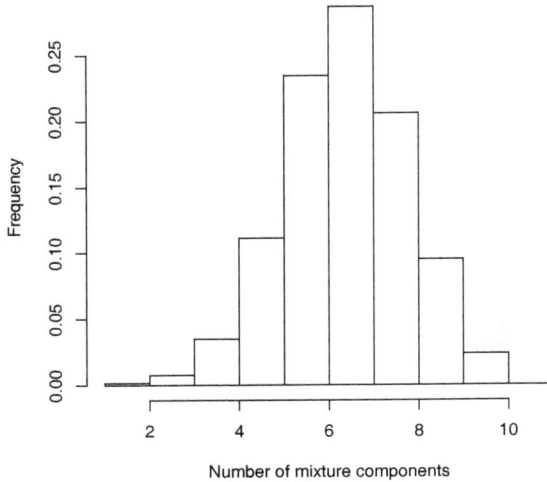

Fig. 5.2.   Galaxy data. Posterior probabilities of the number of components for the Galaxy dataset.

Table 5.2   Galaxy data. KLMCMC results with application of Stephens' relabeling algorithm to the $K = 7$ component model.

Iteration	$\hat{w}_1$	$\hat{w}_2$	$\hat{w}_3$	$\hat{w}_4$	$\hat{w}_5$	$\hat{w}_6$	$\hat{w}_7$
1	0.15	0.13	0.12	0.14	0.17	0.15	0.14
2	0.23	0.08	0.11	0.14	0.30	0.09	0.05
3	0.22	0.09	0.11	0.15	0.29	0.09	0.05
4	0.21	0.09	0.11	0.16	0.29	0.09	0.05
Iteration	$\hat{\mu}_1$	$\hat{\mu}_2$	$\hat{\mu}_3$	$\hat{\mu}_4$	$\hat{\mu}_5$	$\hat{\mu}_6$	$\hat{\mu}_7$
1	21.40	20.31	21.91	20.84	20.97	20.05	21.14
2	22.28	19.14	23.89	21.07	20.54	11.00	30.80
3	22.42	19.06	23.86	21.18	20.43	10.81	30.85
4	22.48	19.04	23.87	21.25	20.40	10.81	30.86
Iteration	$\hat{\sigma}_1^{-2}$	$\hat{\sigma}_2^{-2}$	$\hat{\sigma}_3^{-2}$	$\hat{\sigma}_4^{-2}$	$\hat{\sigma}_5^{-2}$	$\hat{\sigma}_6^{-2}$	$\hat{\sigma}_7^{-2}$
1	1.74	1.74	1.73	1.73	1.74	1.73	1.74
2	1.71	1.80	1.79	1.76	1.71	1.74	1.76
3	1.71	1.81	1.78	1.78	1.70	1.74	1.76
4	1.71	1.81	1.78	1.80	1.69	1.74	1.76

KLMCMC. We ran the KLMCMC program for 10,000 trans-dimensional steps, each containing 100 WinBUGS iterations. We chose $\lambda_b = 700$, $\delta = 0.2$ and $\vartheta = 9.88$, $\mu_0 = median(y)$, $\sigma_0 = range(y)$, $\epsilon = 10 \times \sigma_0^{-2}$, $\alpha = (1, ..., 1)$, $K_{MAX} = 10$ and used the following initial conditions for model parameters:

$$K = 5, \ \beta = 2, \ \sigma_k^{-2} = 1.6 \times 10^{-3} \text{ for } 1 \leq k \leq K. \quad (5.2)$$

The posterior distribution for the number of mixture components is shown in Figure 5.2 with the mode at $K = 7$. Conditional on $K = 7$, the posterior probabilities of the component weights and means for the Galaxy dataset are almost identical for all seven components, as shown in Figures 5.3 and 5.5. To estimate component-specific parameters we used Stephens' relabeling algorithm (Algorithm 3.2); the results are summarized in Table 5.2. After relabeling, the posterior probabilities of the component weights for the Galaxy dataset became distinctly separated. Posterior distributions of the component means are very close to a three-modal distribution

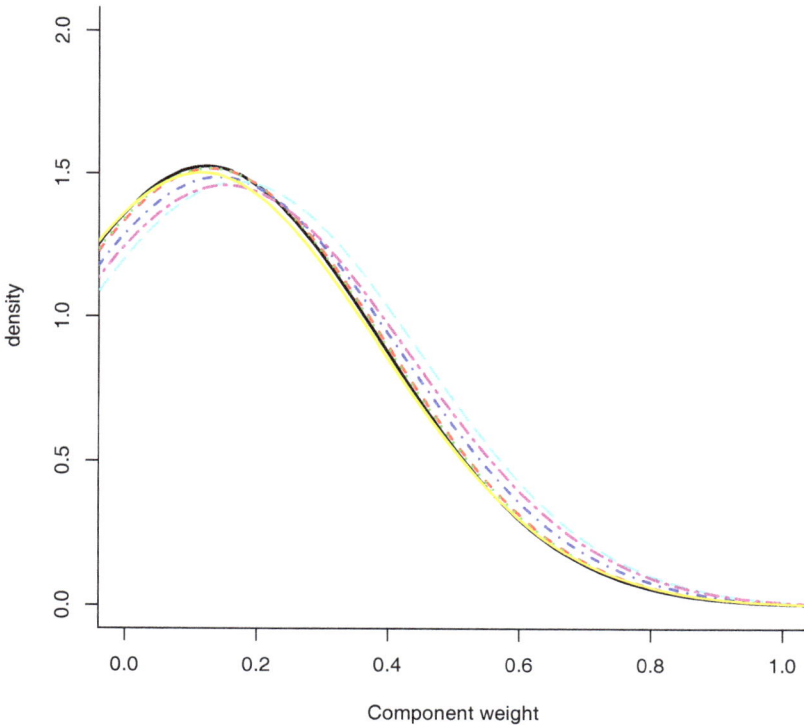

Fig. 5.3. Galaxy data. Posterior probabilities of the component weights for the Galaxy dataset before relabeling.

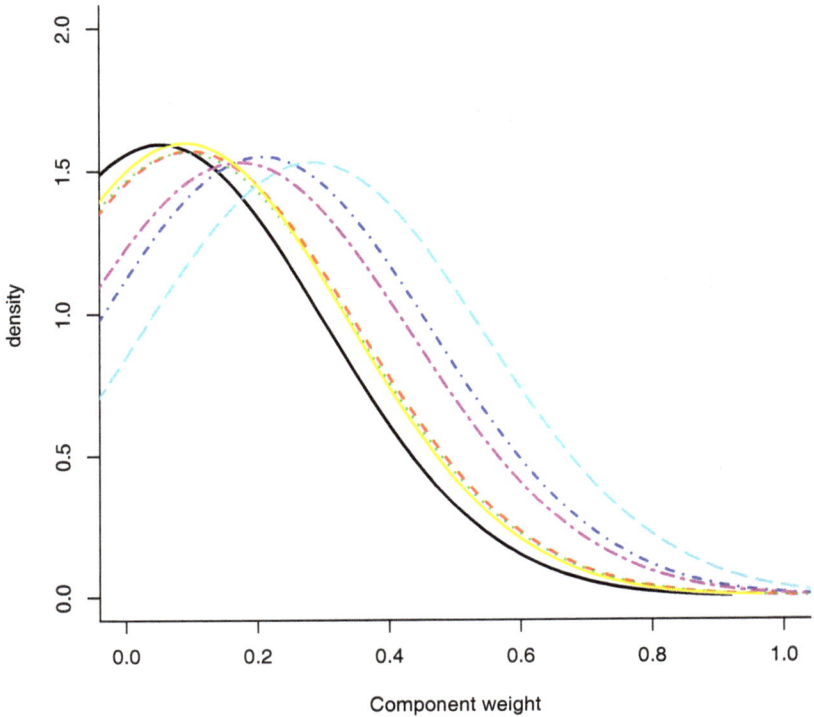

Fig. 5.4. Galaxy data. Posterior probabilities of the component weights for the Galaxy dataset after relabeling.

(Figures 5.3 and 5.5); after relabeling, they separate into seven distinct curves (Figures 5.4 and 5.6). Four iterations of Stephens' relabeling algorithm is enough to successfully deconvolute the mixture of seven distributions.

Reductive stepwise method. We compared the results of the KLM-CMC method to the reductive stepwise method, using BUGS code from Sahu and Cheng (2003). To address the label switching issue, the following re-parametrization was introduced: $\mu_1 < \mu_2 < ... < \mu_K$. We started the simulation with $K = 20$, ran WinBUGS for 20,000 iterations, and discarded the first 10,000 as burn-in. We used the same parameter priors as in the KLMCMC program (Eq. (5.2)). To determine the optimal number of components we considered posterior estimates for the weighted Kullback–

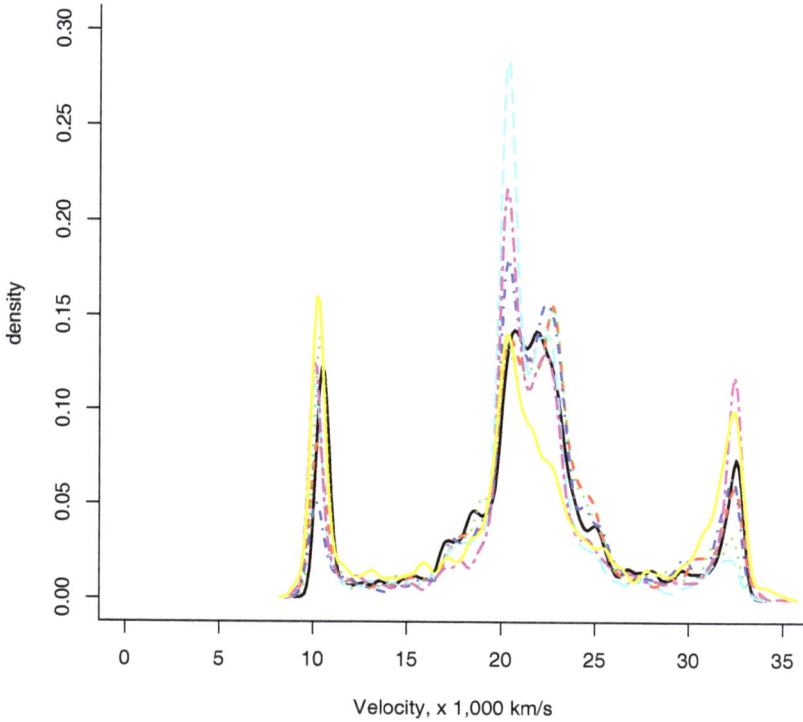

Fig. 5.5.   Galaxy data. Posterior distributions of the component means for the Galaxy dataset before relabeling.

Leibler distance between $K-$ and $(K-1)$-component models and hyper-parameter β (scale parameter of gamma distribution $\sigma_k^{-2} \sim G(\cdot|\vartheta, \beta)$). The relationship between $d(K, K-1)$ and β and the number of mixture components K is shown in Figure 5.7. When $K \geq 8$, the distance $d(K, K-1)$ between $K-$ and $(K-1)$-component models is negligible. For $K < 8$, $d(K, K-1)$ increases as the number of components becomes smaller. A similar trend is observed for $\beta(K)$, which is proportional to the inverse precision of the model.

The weighted Kullback–Leibler distance plot indicates that the cut-off value of $c_k = 0.5$, suggested by Sahu and Cheng (2003), is too high for this dataset. A possible explanation for this phenomenon is that Sahu and Cheng (2003) chose this value from the condition of visual separation

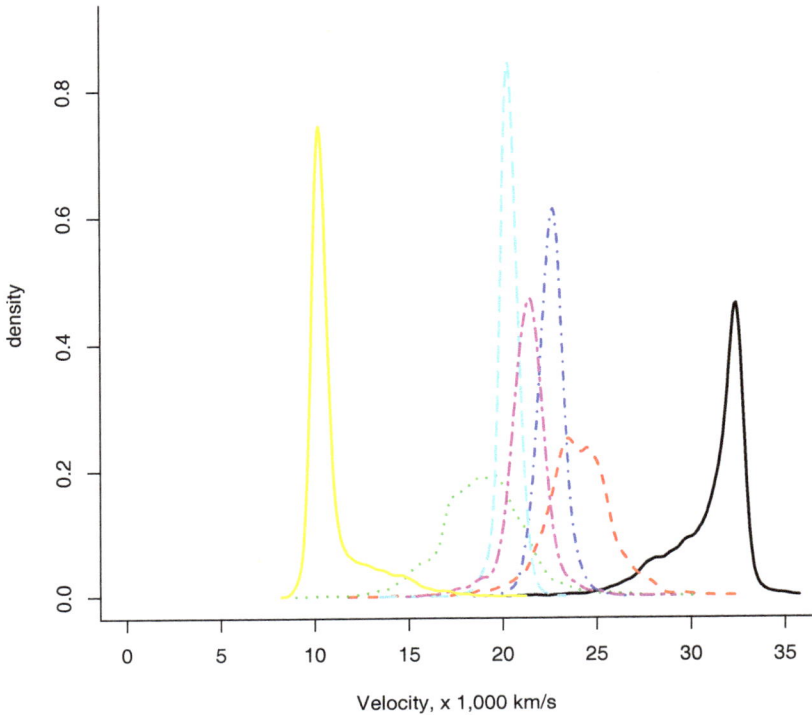

Fig. 5.6.   Galaxy data. Posterior distributions of the component means for the Galaxy
dataset after relabeling.

of two normal distributions and this may not work for larger values of
K. Therefore, we propose an alternative approach: consider the plot of
$d(K, K - 1)$ and select a point K, where "exponential" and "flat" trends
meet. Analysis of the hyper-parameter β can be an additional criterion for
selection of the optimal number of components.

The justification is as follows. From Eq. (5.1), $E[\sigma_k^{-2}|\beta] = \frac{\vartheta}{\beta}$. The model
parameter σ_k^{-2} corresponds to the precision of the kth component. When
the number of components is not sufficient to describe the data, precision is
small and inverse precision $\frac{\beta}{\vartheta}$ is large. When the number of components is
large enough, the precision is also large, and therefore the inverse precision
$\frac{\beta}{\vartheta}$ is small. Since ϑ is a constant, decreasing σ^{-2} corresponds to increasing
β (Figure 5.7, right).

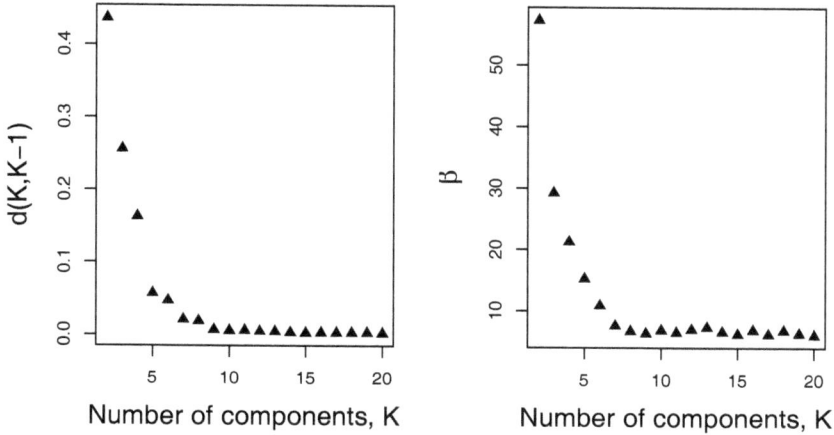

Fig. 5.7. Galaxy data. Determination of the number of components using the reductive stepwise method. Left plot shows weighted KL distance between K- and $(K-1)$-component models; right plot shows inverse precision $\beta(K)$ of the model as a function of the number of model components.

KLMCMC estimated $K = 7$ and the reductive stepwise method estimated $K = 8$ clusters of velocities. Unfortunately, the correct answer is not known. Based on prior analysis of this dataset, between three and seven clusters were reported. Richardson and Green (1997) concluded that the number of components is between five and seven, with six components having the highest probability. Roeder and Wasserman (1997) presented evidence for three velocity clusters. The approach of McLachlan and Peel (2000) supported three components. Stephens (2000b) presented evidence that there are three components for the mixture of normal distributions and four for the mixture of t distributions. More recently, Lau and Green (2007) concluded that the optimal number of clusters is three, and Wang and Dunson (2011) reported five clusters. Therefore, the KLMCMC estimate is within the range of prior estimates and the reductive stepwise method offers one extra component. In the next section we will analyze other benchmark datasets that are closer to Earth and have a defined known answer.

5.2 Simulated Nonlinear Normal Mixture Model

One of the main points of this book is to show how our methods work on nonlinear models. For such models we have to avoid calculating the like-

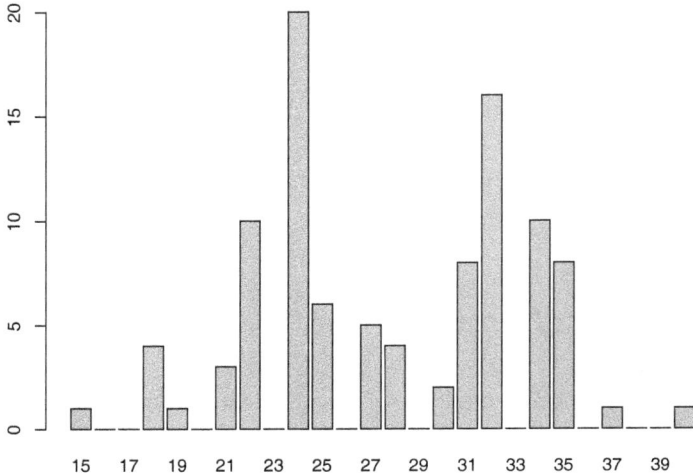

Fig. 5.8.   Dataset generated from nonlinear bimodal mixture distribution.

lihood function as that involves high dimensional integration. The KLM-CMC algorithm is designed for exactly that purpose.

In this example we consider a simulated model with a square root non-linearity. This is the first of a number of nonlinear models to illustrate that KLMCMC works as well on nonlinear models as it does on linear models. In this case, the KLMCMC method is applied to a simulated nonlinear model with a square root nonlinearity.

We generated $N = 100$ data points (Figure 5.8) from the nonlinear mixture distribution:

$$y_i \sim N(\cdot | \sqrt{|\theta_i|}, \tau_e^{-1}),$$

$$\theta_i \sim N(\cdot | \lambda_1, \tau^{-1}) \text{ , for } 1 \leq i \leq 50, \qquad (5.3)$$

$$\theta_i \sim N(\cdot | \lambda_2, \tau^{-1}) \text{ , for } 51 \leq i \leq 100.$$

The values of λ_1 and λ_2 were generated from the normal distributions $N(\cdot | 500, 100)$ and $N(\cdot | 1000, 100)$, respectively. "Precision" parameters were $\tau = 0.01$ and $\tau_e = 0.1$. We assumed that the parameters τ and τ_e are known and our task was to find the number of mixture components, estimate values of $\lambda_{1,2}$ and component weights $\pi_{1,2}$ from the "observed" data. In the KLMCMC calculation, we used the following model:

$$y_i \sim N(\cdot | \sqrt{\theta_i}, \tau_e^{-1}),$$

$$\theta_i \sim \sum_{k=1}^{K} w_k N(\cdot | \lambda_k, \tau^{-1}),$$

$$\pi_k \sim Dir(\cdot | \alpha),$$

$$\lambda_k \sim N(\cdot | \lambda_0, \tau_0^{-1}),$$

(5.4)

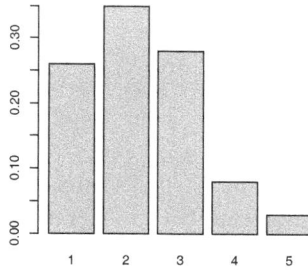

Fig. 5.9.   Nonlinear bimodal dataset: Distribution of the number of component K.

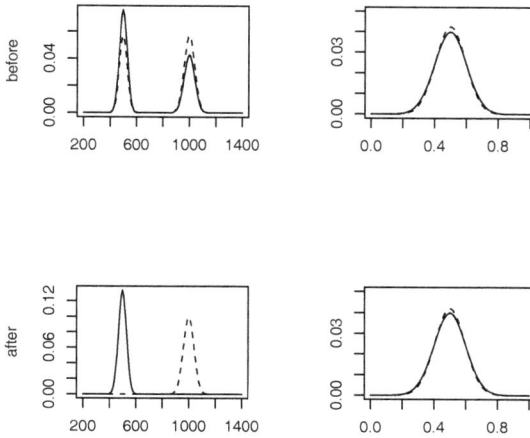

Fig. 5.10.   Nonlinear bimodal dataset, $K = 2$: Output of KLMCMC for component means (left) and component weights (right) before and after relabeling.

Table 5.3 Results of KLMCMC after relabeling for the simulated nonlinear dataset.

N	$\hat{\pi}_1$	$\hat{\pi}_2$	$\hat{\mu}_1$	$\hat{\mu}_2$	$\hat{\sigma}_1$	$\hat{\sigma}_2$
1	0.503	0.469	739.83	775.62	10.00	10.00
2	0.494	0.505	501.23	1008.23	10.01	10.01
3	0.493	0.506	500.39	1007.76	10.01	10.01

where

$$\lambda_0 = 750,$$

$$\tau_0 = 1.0 \times 10^{-5}, \ \tau = 0.01, \ \tau_e = 0.1, \tag{5.5}$$

$$\alpha_k = 1 \text{ for } k = 1, ..., K.$$

Initial conditions for parameters λ_k were set to be $\lambda_k = \lambda_0$ for all k. We applied the KLMCMC method and for the birth–death step of the algorithm we chose $\lambda_b = 0.9$. This choice helps the number of components to "oscillate" around the $K = 2$ state. To improve convergence, we also used the RPS (Algorithm 3.1) at each birth–death step [Fruhwirth-Schnatter (2001)]. Due to the nonlinearity, this model required longer burn-in time, as compared to the Eyes and Galaxy model. At each step, we let the Gibbs sampler run for 4,000 iterations and then used 100 points to estimate model parameters.

As before, the WinBUGS output for π and λ were used as the initial values for the next step. We made 2,500 iterations of the KLMCMC algorithm and achieved acceptable mixing, bimodal for λ and unimodal for π. The distribution of the number of components peaked at $K = 2$. The distributions of K, π and λ are given in Figures 5.9 and 5.10.

We applied the Stephens' Relabeling algorithm to these results and the output is shown in Table 5.3 and in Figure 5.10. From Table 5.3 and Figure 5.10, we can conclude that KLMCMC followed by relabeling has correctly recovered model parameters.

5.3 Linear Normal Mixture Model: Boys and Girls

We are going to illustrate methods described in previous sections using the well-known dataset presented by Potthoff and Roy (1964). Researchers at the Dental School at the University of North Carolina conducted a study of dental measurements ("dental measurements of the distance from the center of the pituitary gland to the pterygomaxillary tissue") for a group

Table 5.4 Boys and Girls dataset [Potthoff and Roy (1964)], containing dental measurements for 27 children.

ID	Gender	8 years	10 years	12 years	14 years
1	boy	26	25	29	31
2	boy	21.5	22.5	23	26.5
3	boy	23	22.5	24	27.5
4	boy	25.5	27.5	26.5	27
5	boy	20	23.5	22.5	26
6	boy	24.5	25.5	27	28.5
7	boy	22	22	24.5	26.5
8	boy	24	21.5	24.5	25.5
9	boy	23	20.5	31	26
10	boy	27.5	28	31	31.5
11	boy	23	23	23.5	25
12	boy	21.5	23.5	24	28
13	boy	17	24.5	26	29.5
14	boy	22.5	25.5	25.5	26
15	boy	23	24.5	26	30
16	boy	22	21.5	23.5	25
17	girl	21	20	21.5	23
18	girl	21	21.5	24	25.5
19	girl	20.5	24	24.5	26
20	girl	23.5	24.5	25	26.5
21	girl	21.5	23	22.5	23.5
22	girl	20	21	21	22.5
23	girl	21.5	22.5	23	25
24	girl	23	23	23.5	24
25	girl	20	21	22	21.5
26	girl	16.5	19	19	19.5
27	girl	24.5	25	28	28

of 27 boys and girls. Dental measurements were collected at the ages of 8, 10, 12 and 14. The dataset is shown in Table 5.4 and Figure 5.11; it was previously analyzed by Wakefield *et al.* (1994).

From a visual inspection of the data, it appears that boys have a higher growth rate than girls. We aim to scrutinize this observation to achieve a certain degree of confidence. Let $j = 1, ..., 4$ index over time points, $x_j = \{8, 10, 12, 14\}$ years of measurement, $i = 1, ..., 27$ index over children; children 1–16 are boys and 17–27 are girls. The variables y_{ij} represent the measurements of child i at time x_j.

First stage model:

$$p(y_{ij}|\theta_i, \tau) = N(y_{ij}|\alpha_i + x_j\beta_i, \tau^{-1}),$$

$$i = \{1, ..., 27\} \text{ and } j = \{8, 10, 12, 14\}, \tag{5.6}$$

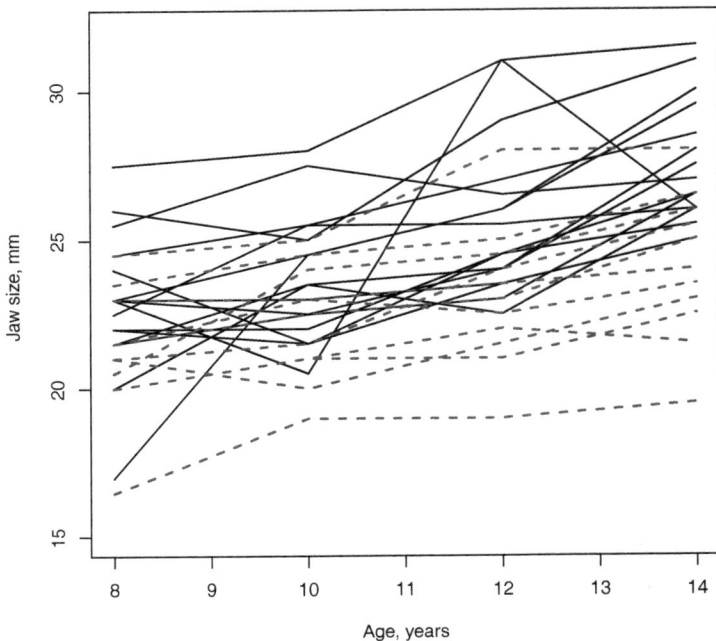

Fig. 5.11. Boys and Girls. Dental measurements of the distance from the center of the pituitary gland to the pterygomaxillary tissue. Boys' measurements are solid lines and girls' are dashed lines. Children's age in years is plotted on the X axis, corresponding dental measurements in millimeters on the Y axis.

where the random variable $\theta_i = (\alpha_i, \beta_i)$ is a two-dimensional vector; coefficients α_i and β_i are intercepts and slopes for the ith child growth curve. The number of mixture components is $K = 2$.

Second stage model:

$p(\theta_i, \phi)$ is a $(K = 2)$-component function $F^{(2)}$, and

$$F^{(2)}(\theta_i, \phi) = \sum_{k=1}^{2} w_k N(\theta_i | \mu^k, \Sigma^k). \qquad (5.7)$$

Third stage model:

$$p(\tau) = G(\tau | \frac{\nu_0}{2}, \frac{\nu_0 \tau_0}{2}),$$

$$p(\phi^k) = p(\mu_k) p(\Sigma_k^{-1}) = N(\mu_k | \eta, C) W(\Sigma_k^{-1} | (\rho R \rho)^{-1}, \rho), k \in \{1, 2\}. \qquad (5.8)$$

A conjugate prior for component weight vector is $w \sim Dirichlet(\cdot | \alpha)$.

Reductive stepwise method. Since the scientific question is whether there is a difference between jaw growth in boys and girls, we need to compare two- and one-component models. For every child i, the two-component and collapsed one-component mixtures are given by

$$F^{(2)}(\theta_i|.) = \sum_{k=1}^{2} w_k F_k(\theta_i|\mu_k, \Sigma_k) = \sum_{k=1}^{2} w_k N(\theta_i|\mu_k, \Sigma_k),$$

$$(5.9)$$

$$F_{12}^{*(1)}(\theta_i|\mu^*, \Sigma^*) = N(\theta_i|\mu^*, \Sigma^*).$$

To decide between one- and two-component mixtures, assume that we have successfully fitted the two-component model and obtained distributions of all parameters. In this case, $w_1 + w_2 = 1$ and the weighted Kullback–Leibler distance d^* is equal to

$$d^*(F^{(2)}, F_{12}^{*(1)}) = \sum_{k \in \{1,2\}} \frac{w_k}{2} log(det(\Sigma^*(\Sigma_k)^{-1}))$$

$$+ \sum_{k \in \{1,2\}} \frac{w_k}{2} trace((\Sigma^*)^{-1}\Sigma_k) + (\mu^* - \mu_k)^T(\Sigma^*)^{-1}(\mu^* - \mu_k) - 1, \quad (5.10)$$

where, for $K = 2$,

$$\mu^* = w_1\mu_1 + w_2\mu_2,$$

$$\Sigma^* = w_1\Sigma_1 + w_2\Sigma_2 + w_1 w_2 (\mu_1 - \mu_2)(\mu_1 - \mu_2)^T. \quad (5.11)$$

Substituting the expression for μ^* into the formula for the minimum weighted Kullback–Leibler distance $d^*(F^{(2)}, F_{12}^{*(1)})$, we get

$$d^*(F^{(2)}, F_{12}^{*(1)}) = \sum_{k \in \{1,2\}} \frac{w_k}{2} \{trace((\Sigma^*)^{-1}\Sigma_k) + log(det(\Sigma^*(\Sigma_k)^{-1})) - 2\}$$

$$+ \frac{w_1 w_2}{2}(\mu_1 - \mu_2)^T(\Sigma^*)^{-1}(\mu_1 - \mu_2). \quad (5.12)$$

To deal with possible label switching, we re-parametrize μ_2 as $\mu_1 + \delta$, where δ is a two-dimensional vector with positive components. We accept the one-component model if distance $d(F^{(2)}, F_{12}^{*(1)})$ is small. In the Bayesian approach, it is necessary to calculate posterior probability:

$$P_{c_2}(2) = P(d^*(F^{(2)}, F_{12}^{*(1)}) \leq c_2|data). \quad (5.13)$$

If $P_{c_2}(2) > \alpha$, we should choose the one-component model. We implemented this model in the WinBUGS framework (Eq. (D.1)), for simplicity, assuming equal variances for the boys' and girls' components $\Sigma_1 = \Sigma_2$. The WinBUGS code is in the Appendix D.1. At every iteration we computed the distance d between two- and one-component models. If $P(d < c_2)$ is large, then the one-component model is adequate. Using the results of Sahu and Cheng (2003) we used $c_2 = 0.5$. Following Richardson and Green (1997) and Zhou (2004), we used the least-squares estimates to define weakly informative priors. The values for the hyper-parameters are $\nu_0 = 0.001$, $\tau_0 = 1$, $\rho = 2$, $Z_1 = 1$, $Z_{26} = 2$,

$$C^{-1} = \begin{pmatrix} 10^{-6} & 0 \\ 0 & 10^{-6} \end{pmatrix}, R = \begin{pmatrix} 1 & 0 \\ 0 & 0.1 \end{pmatrix},$$

$$\eta = \left(23.84, 0.69 \right).$$

Initial values were set to $\mu_{1,2} = \{23.84, 0.69\}$, $\tau = 1$. The Gibbs sampler was run for 200,000 iterations and the first 100,000 burn-in iterations were discarded. Results are shown in Tables 5.5, 5.6 and Figure 5.12.

In Table 5.5, d corresponds to the posterior estimate of the weighted Kullback–Leibler distance between two- and one-component models; w_k is a probability to be assigned to the kth component; $\mu = (\mu_1^i, \mu_2^i)$ is the mean vector of the kth component; σ_{ij} $1 \le i, j \le 2$ are the elements of the covariance matrix; $\mu^* = (\mu^{1*}, \mu^{2*})$ is the mean of the collapsed version and σ_{ij}^* are the elements of the covariance matrix for the collapsed version. Posterior probability for $c_2 = 0.5$ is $P(d^*(F^{(2)}, F_{1,2}^{*(1)}) \le 0.5|data) = 0.44 < 0.5$, and we do not accept the one component model.

As a result of the reductive stepwise method, we assigned 62% of children to the "boys" component and 38% to the "girls" component. The growth curve for the boys is $y_b(AGE) = 24.78 + 0.70 \times (AGE - 8)$, and for the girls is $y_g(AGE) = 22.34 + 0.64 \times (AGE - 8)$.

As an alternative to the re-parametrization, we consider approach of Stephens (2000b) as a post-processing step for the reductive stepwise method (see Table 5.7). Following Zhou (2004), we use informative priors for parameters $\nu_0 = 0$, $\tau_0 = 1$, $\rho = 2$, $C^{-1} = 0$,

$$R = \begin{pmatrix} 1 & 0 \\ 0 & 0.1 \end{pmatrix},$$

$$\eta = \left(23.84, 0.69 \right).$$

As can be seen from Figure 5.12, the posterior distributions of the intercepts of the jaw growth curves (solid curves) are bimodal. This is an

Table 5.5 Boys and Girls dataset. Reductive stepwise method. Parameters estimated by mixture model with two components and distances between components.

Node	Mean	SD	MC error	2.5%	Median	97.5%
$\mu_1^1 = E(\alpha_1)$	24.78	0.7	0.01	23.5	25.2	29.8
$\mu_1^2 = E(\beta_1)$	0.7	0.16	0.004	0.05	0.7	1.7
$\mu_2^1 = E(\alpha_2)$	22.34	1.04	0.02	17.44	22.7	24.1
$\mu_2^2 = E(\beta_2)$	0.64	0.2	0.006	−0.5	0.6578	1.6
μ^{1*}	23.55	2.42	0.21	22.73	23.83	24.74
μ^{2*}	0.7	0.24	0.004	0.33	0.7	1.06
σ_{11}^*	4.12	2.27	0.08	1.33	3.7	9.1
σ_{12}^*	0.13	0.5	0.02	−0.7	0.1	1.07
σ_{22}^*	0.5	0.9	0.04	0.2	0.5	1.06
σ_{11}	2.04	1.5	0.02	0.16	1.8	5.7
σ_{12}	0.05	0.2	0.003	−0.48	0.04	0.7
σ_{22}	0.05	0.03	5E-4	0.23	0.4	0.8
d	1.7987	0.6	0.03	0.01	0.68	2.4
w_1	0.62	0.22	0.02	0.06	0.63	0.93
w_2	0.38	0.16	0.02	0.03	0.41	0.61

Table 5.6 Boys and Girls dataset. Reductive stepwise method. Posterior parameter values with component labels removed. Sample size 100,001, burn-in 100,000 iterations.

Node	Mean	SD	MC error	2.5%	Median	97.5%
w_1	0.5044	0.0971	0.001414	0.3075	0.508	0.683
w_2	0.4956	0.0971	0.001414	0.317	0.492	0.6925
$\Sigma_{1,1}^{-1}$	0.8406	0.8637	0.01613	0.1813	0.5745	3.179
$\Sigma_{1,2}^{-1}$	−0.6527	3.504	0.06738	−8.28	−0.541	6.59
$\Sigma_{2,2}^{-1}$	40.57	24.93	0.322	9.549	34.77	104.4
μ_1^1	23.75	1.099	0.03769	21.81	23.67	25.86
μ_2^1	0.6951	0.1695	0.003803	0.3562	0.6961	1.029
μ_1^2	24.07	1.814	0.0681	20.7	24.14	27.44
μ_2^2	0.7045	0.2391	0.006815	0.2363	0.7036	1.184
τ	0.1711	0.0287	1.91E−4	0.1197	0.1694	0.232

indication of label switching. We applied Stephens' algorithm [Stephens (2000b)] (Algorithm (3.2) in this book) for decoupling. In Figure 5.12 dotted unimodal curves show distributions of the values of intercepts obtained as a result of this algorithm.

The mean time curves for boys and girls are $y_b(AGE) = 24.95 + 0.82 \times (AGE − 8)$ and $y_g(AGE) = 22.83 + 0.58 \times (AGE − 8)$. The proportion of boys was estimated as 0.61 and the proportion of girls as 0.39.

KLMCMC. We also analyzed this dataset using the KLMCMC method followed by the random permutation and relabeling steps. For

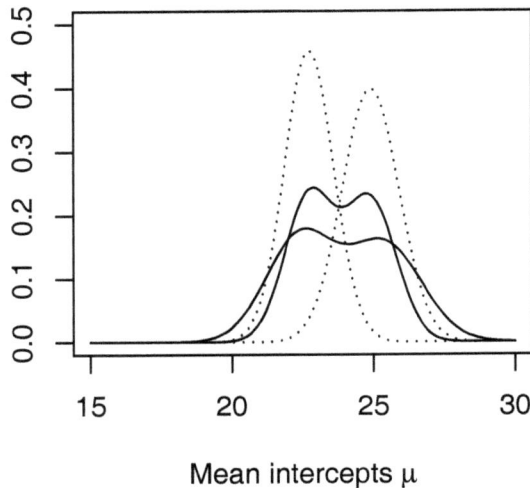

Mean intercepts μ

Fig. 5.12.   Boys and Girls dataset. Reductive stepwise method. Solid lines show posterior distributions of mean intercepts for the two categories. Dotted lines show the result of the application of Stephens' relabeling algorithm to these distributions.

Table 5.7   Boys and Girls dataset. Results of application of Stephens' algorithm to the output of the reductive stepwise method. Stable point was reached in ten iterations of the algorithm. w_1 and w_2 are weights of the boys and girls components, and $\mu_{(1,2)}^{(1,2)}$ are estimates of intercept (1) and slope (2) for the dental growth curve for boys (1) and girls (2).

Iteration	w_1	w_2	μ_1^1	μ_2^1	μ_1^2	μ_2^2
1	0.51704	0.48296	24.0356	23.7418	0.712364	0.689615
2	0.51709	0.48291	24.2866	23.5893	0.732944	0.676166
3	0.517312	0.482688	24.5252	23.4218	0.768371	0.651566
4	0.518422	0.481578	24.7781	23.1992	0.778732	0.643464
5	0.523559	0.476441	24.9538	22.9439	0.789309	0.634383
6	0.545181	0.454819	24.9638	22.8517	0.799694	0.624228
7	0.582752	0.417248	24.9568	22.8373	0.809404	0.612581
8	0.604409	0.395591	24.9548	22.8346	0.816269	0.583509
9	0.604696	0.395304	24.9544	22.8341	0.816206	0.583431
10	0.614693	0.385307	24.9543	22.8339	0.816174	0.583392

the KLMCMC algorithm, we used the following parameters: $B = 2$, fixed $\tau = 0.17$ (as determined by the reductive stepwise method), $\rho = 2$,

$$R = \begin{pmatrix} 0.5 & 0 \\ 0 & 5 \end{pmatrix},$$

$$C = \begin{pmatrix} 10^{-3} & 0 \\ 0 & 10 \end{pmatrix},$$

$$\eta = (23.84, 0.69).$$

We ran the KLMCMC algorithm for 400 birth–death iterations, each containing 1,000 of the Gibbs sampler steps. We sampled 22,700 iterations from this output. Results of the application of the KLMCMC algorithm are presented in Figures 5.13, 5.14, 5.17, 5.16, and 5.17.

KLMCMC estimates of the jaw growth curves are $y_b(AGE) = 24.28 + 0.70 \times (AGE - 8)$ and $y_g(AGE) = 23.84 + 0.67 \times (AGE - 8)$. The proportion of boys was estimated as 0.61 and the proportion of girls as 0.39. Out of 11 girls, one has a boy-like jaw growth curve $y(AGE) = 24.35 + 0.69 \times (AGE - 8)$. She was labeled as a boy by KLMCMC as well as the reductive stepwise method. Both methods predict that boys have larger intercepts and slopes of the growth curves and agree on component weights. However, both implementations of the reductive stepwise method predict slightly larger differences between boys' and girls' intercepts than the KLMCMC method (11% vs. 2% difference). Since we did not accept the one-component model, our statistical analysis confirms popular public belief that boys' jaws grow faster and become bigger than girls' jaws. The two methods of dealing with label-switching problems, relabeling and re-parametrization, produced consistent results for posterior estimates of the model parameter values.

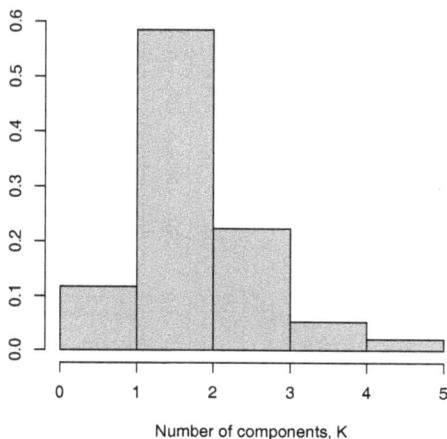

Fig. 5.13. Boys and Girls dataset. Distribution of the number of components analyzed by the KLMCMC method.

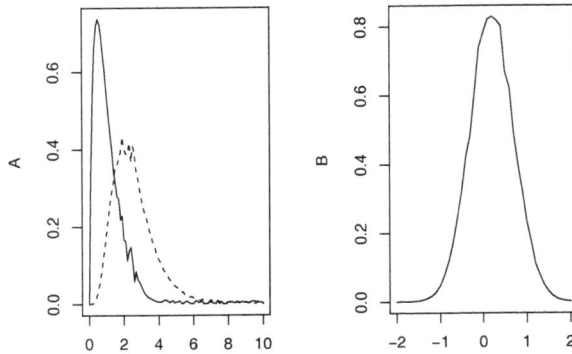

Fig. 5.14.   Boys and Girls dataset. Posterior distributions of components of the inverse covariance matrix, diagonal elements (A: solid line Σ_{11}^{-2}, dashed line Σ_{22}^{-2}) and the off-diagonal elements $\Sigma_{12}^{-2} = \Sigma_{21}^{-2}$ (B), analyzed by the KLMCMC method.

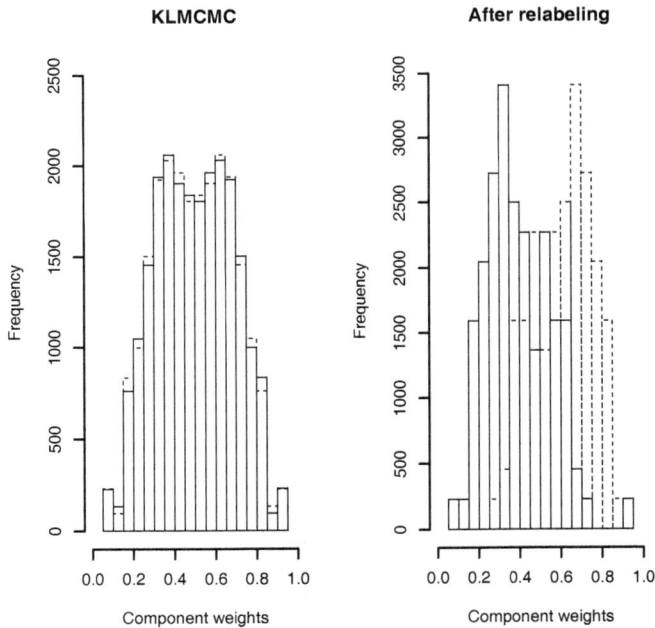

Fig. 5.15.   Boys and Girls dataset. Solid lines — girls. Dashed lines — boys. Posterior distributions of component weights, analyzed by the KLMCMC method, and conditioned on $K = 2$. Distributions are shown before (left) and after (right) Stephens' Relabeling.

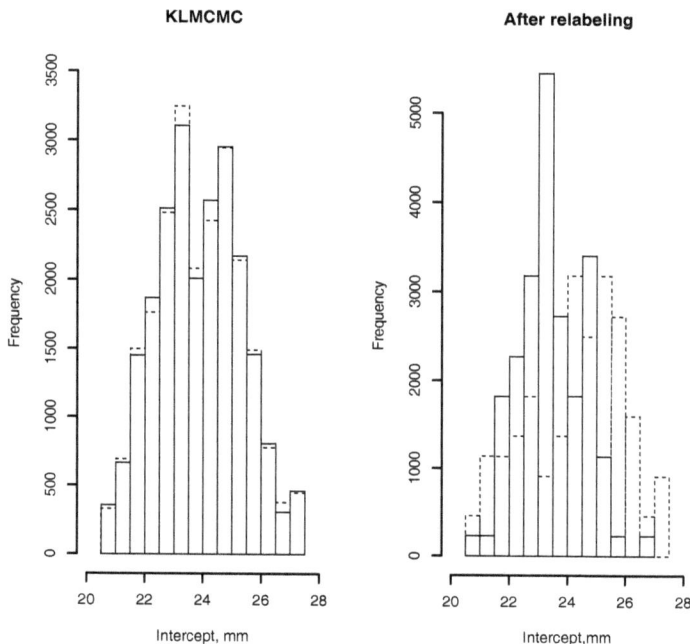

Fig. 5.16.   Boys and Girls dataset. Solid lines — girls. Dashed lines — boys. Posterior distributions of intercep, analyzed by the KLMCMC method, and conditioned on $K = 2$. Distributions are shown before (left) and after (right) Stephens' Relabeling.

5.4   Nonlinear Pharmacokinetics Model and Selection of Prior Distributions

In this section, we will apply the KLMCMC method to the simulated pharmacokinetics (PK) one-compartment model, described earlier in Section 2.2.4 and Eq. (2.30).

$$C(t) = \frac{D}{V} exp(-\kappa t), \tag{5.14}$$

where D is the administered dosage, κ is the patient-specific elimination constant, and V is a volume of distribution that is specific to the drug as well as to the patient. Drugs that are highly lipid soluble have a very high volume of distribution and those that are lipid insoluble remain in the blood and have a low V.

If we consider administering a drug to a group of N subjects and taking a series of T measurements of the drug level in the blood, the measured

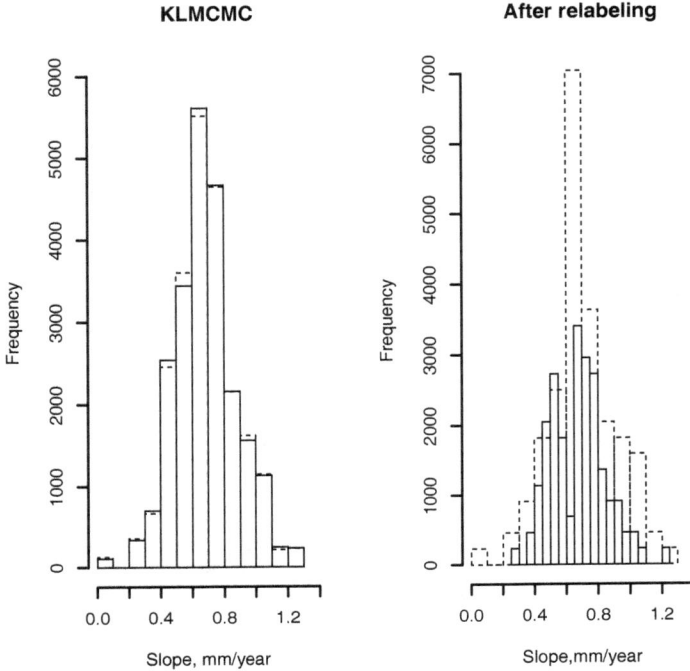

Fig. 5.17. Boys and Girls dataset. Solid lines — girls. Dashed lines — boys. Posterior distributions of slope, analyzed by the KLMCMC method, and conditioned on $K = 2$. Distributions are shown before (left) and after (right) Stephens' Relabeling.

levels can be described by the following PK model:

$$Y_{ij} = \frac{D}{V_i} exp(-\kappa_i t_j) + \sigma_e e_{ij}, \tag{5.15}$$

where Y_{ij} is a measurement for the ith subject at time t_j, for $i = 1, ..., N$ and $j = 1, ..., T$. Further, measurement errors e_{ij} are assumed to have independent normal $N(0, 1)$ distributions. The patient-specific elimination constants κ_i can be described as a mixture of K normal distributions:

$$\kappa_i \sim \sum_{k=1}^{K} w_k N(\cdot | \mu_k, \sigma_k^2). \tag{5.16}$$

The patient-specific volume variables V_i can be described by a single normal distribution $V_i \sim N(\cdot | V_0, \sigma_v^2)$. For $K = 2$, this reflects the nature of fast and slow acetylators (fast and slow acetylators differ in inherited (genetic) ability to metabolize certain drugs). Therefore, the resulting model is as follows:

$$Y_{ij} = \frac{D}{V_i} exp(-\kappa_i t_j) + \sigma_e e_{ij}, \text{ for } i = 1, ..., N, \, j = 1, ..., T,$$

$$\kappa_i \sim \sum_{k=1}^{2} w_k N(\cdot|\mu_k, \sigma_k^2),$$

$$\mu_k \sim N(\cdot|\mu_0, \sigma_0^2), \text{ for } k = 1, ..., K, \tag{5.17}$$

$$w \sim Dir(\cdot|\alpha), \text{ where all } \alpha_k = 1,$$

$$V_i \sim N(\cdot|V_0, \sigma_v^2).$$

This model can be used to describe the bimodal nature of certain drugs under IV administration. In this case $K = 2$, and we simulated "observations" from the model in Eq. (5.17) using the following "truth" model:

$$K = 2, \, T = 5, \, N = 100, \, \sigma_e^2 = 0.01, \, D = 100,$$
$$\mu_1 = 0.3, \, \mu_2 = 0.6, \, \sigma_1 = \sigma_2 = 0.06,$$
$$w_1 = 0.8, \, w_2 = 0.2, \, V_0 = 20, \, \sigma_v = 2.$$

We have generated time-series "observations" for $N = 100$ patients. Time points t_j, $j = 1, ..., 5$ are assumed to be $(1.5, 2, 3, 4, 5.5)$. The population distributions of the generated values of κ_i and V_i are shown in Figure 5.18 and the "observations" are in Figure 5.19. Mean volume of distribution V is 20.08, standard deviation of V is 2.06, mean elimination constant κ is 0.38 and standard deviation of κ is 0.13. In a general case, we are going to use the PK model in Eq. (5.17) for this purpose.

We will use KLMCMC followed by Stephens' relabeling algorithm to find population parameters. There are several possible ways to model the parameters.

Case 1

In the first case we assume equality of variances of the components $\sigma_1 = ... = \sigma_K = \sigma$; the volume of distribution is set to be constant and the same for all patients, $V_i \equiv 20$. Therefore, the following equation should be added to our general model in Eq. (5.17), namely:

$$\sigma^{-2} \sim G(\cdot|\xi, \beta). \tag{5.18}$$

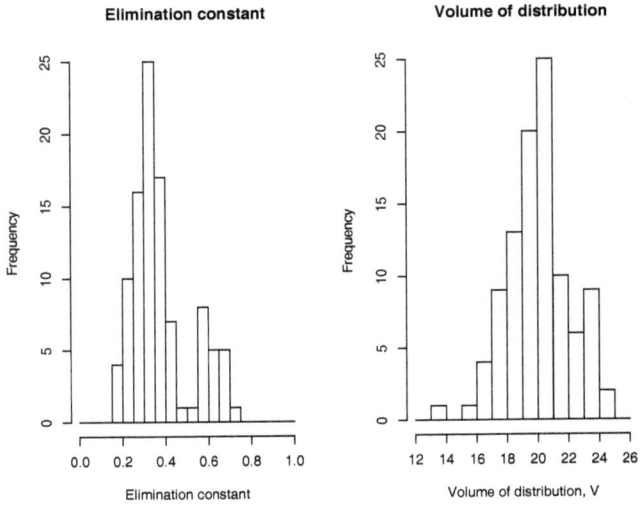

Fig. 5.18.   One-compartment PK model. Population distribution of the generated values of elimination constant κ and volume of distribution V.

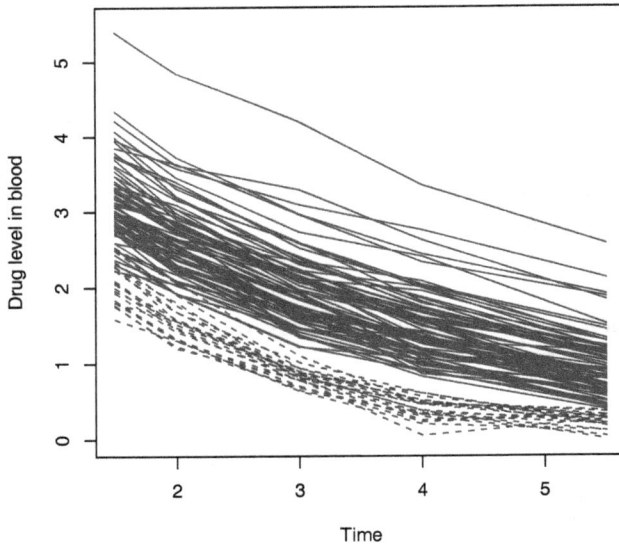

Fig. 5.19.   One-compartment PK model. Generated values of Y_{ij}; solid lines correspond to $\mu_1 = 0.3$ and dashed lines correspond to $\mu_2 = 0.6$.

Table 5.8 One-compartment PK model, Case 1. Results of application of Stephens' relabeling algorithm to the $K = 2$ model, with randomization.

Iter	$\hat{w}_1$	$\hat{w}_2$	$\hat{\mu}_1$	$\hat{\mu}_2$	$\hat{\sigma}$
1	0.501	0.499	0.358	0.358	0.07
2	0.793	0.207	0.319	0.600	0.07

To approximate the hyper-parameters of prior distributions we used nonlinear least-squares estimation. The resulting parameter values were $\lambda = 0.356$, $\sigma_0 = 1$. Thus, the resulting parameter values were $\xi = 30$, and $\beta = 0.1$. Using KLMCMC with RPS, we assumed the Poisson birth rate to be $\lambda_B = 2$ and made 500 KLMCMC steps (each including 1,000 WinBUGS iterations), obtaining the results shown in Figure 5.20 and Table 5.8. The KLMCMC method was able to recover the correct parameter values for $(\mu_1, \mu_2, \sigma, w_1, w_2)$.

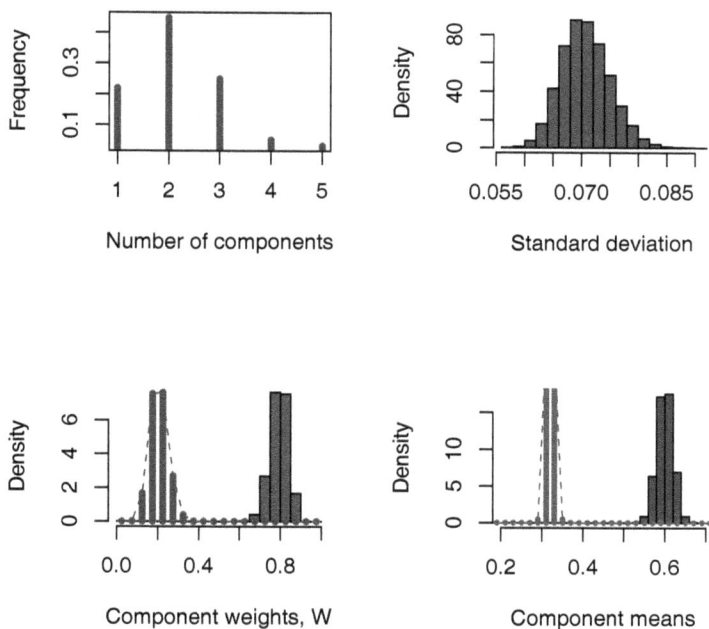

Fig. 5.20. One-compartment PK model, Case 1. Posterior distribution of the number of components (top left), standard deviation σ (top right); Conditional on $K = 2$: component weights (bottom left), component means (bottom right).

Model parameters w and μ were correctly estimated by the algorithm based on posterior means. Since we incorrectly assumed that $V_i = 20$ for all $i = 1, ..., 100$, the standard deviation σ was estimated to be higher than the corresponding population parameter.

Case 2

Let us make a more realistic assumption that the patient-specific volume V_i is normally distributed with known mean and standard deviation:a

$$V_i \sim N(\cdot | V_0, \sigma_v^2), \text{ for } i = 1, ..., N. \tag{5.19}$$

We assume that the known model parameters are

$$V_0 = 20, \ \sigma_v = 2, \ \sigma_e = 0.1 \tag{5.20}$$

This condition in Eq. (5.19) was appended to our general model in Eq. (5.17) and Eq. (5.18).

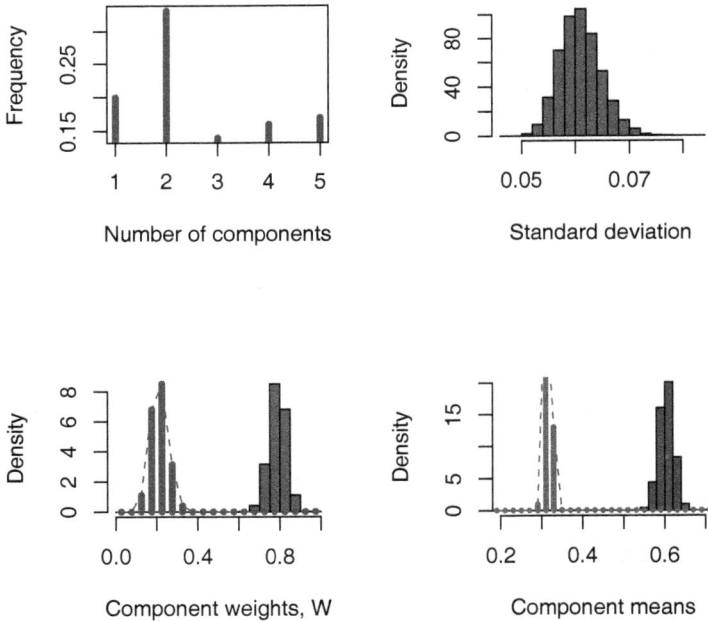

Fig. 5.21.   One-compartment PK model, Case 2. Posterior distribution of the number of components (top left), standard deviation σ (top right) ; Conditional on $K = 2$: component weights (bottom left), component means (bottom right).

Table 5.9 One compartment PK model, Case 2. Results of application of Stephens' relabeling algorithm to the $K = 2$ model, variable V_i.

Iter	$\hat{w}_1$	$\hat{w}_2$	$\hat{\mu}_1$	$\hat{\mu}_2$	$\hat{\sigma}$
1	0.50	0.50	0.35	0.35	0.061
2	0.79	0.21	0.32	0.60	0.061

The unknown parameters were (K, μ_k, σ, w_k). We ran the KLMCMC with the RPS for 500 steps (each including 1,000 WinBUGS iterations), and applied Stephens' relabeling as a post-processing step. The posterior distributions of the model parameters after relabeling are shown in Figure 5.21. Table 5.9 shows the output of Stephens' relabeling algorithm for $K = 2$.

Case 3

A more realistic problem arises when we generate the values for V from a unimodal normal distribution, but "forget" about it and assume the volume of distribution V is described by a mixture of normals $V \sim \sum_k w_k N(\cdot | v_k, (\sigma_k^v)^2)$. We also remove the condition of homoscedacity (equality of variances) σ_k^2. In this case, the equation for the elimination constant should be modified, and additional conditions are appended to the general model in Eq. (5.17):

$$V_i \sim \sum_k^K w_k N(\cdot | v_k, (\sigma_k^v)^2), \text{ for } k = 1, ..., K,$$

$$\kappa_k \sim \sum_k^K w_k N(\cdot | \mu_k, \sigma_k^2), \ v_k \sim N(\cdot | V_0, (\sigma_0^v)^2), \qquad (5.21)$$

$$\frac{1}{(\sigma_k)^2} \sim G(\cdot | \xi, \beta), \ \frac{1}{(\sigma_k^v)^2} \sim G(\cdot | \xi_v, \beta_v).$$

We selected parameters of prior distributions using the logic of the simple model (Eq. (5.21)) combined with the "trial and error" approach: we made multiple short runs of the model in WinBUGS, selecting the least informative values of parameters of prior distributions for which the Gibbs sampler did not diverge. For each parameter, we recorded the value for which the transition between proper and improper posterior distributions occurs and used this value in the KLMCMC run. Consider, for example, parameter σ_v; $(\sigma_v)^{-2}$ has a $G(\xi_v, \beta)$ distribution; therefore its standard deviation is $\sqrt{\xi_v}/\beta$. We plotted values of the posterior estimate of the inverse

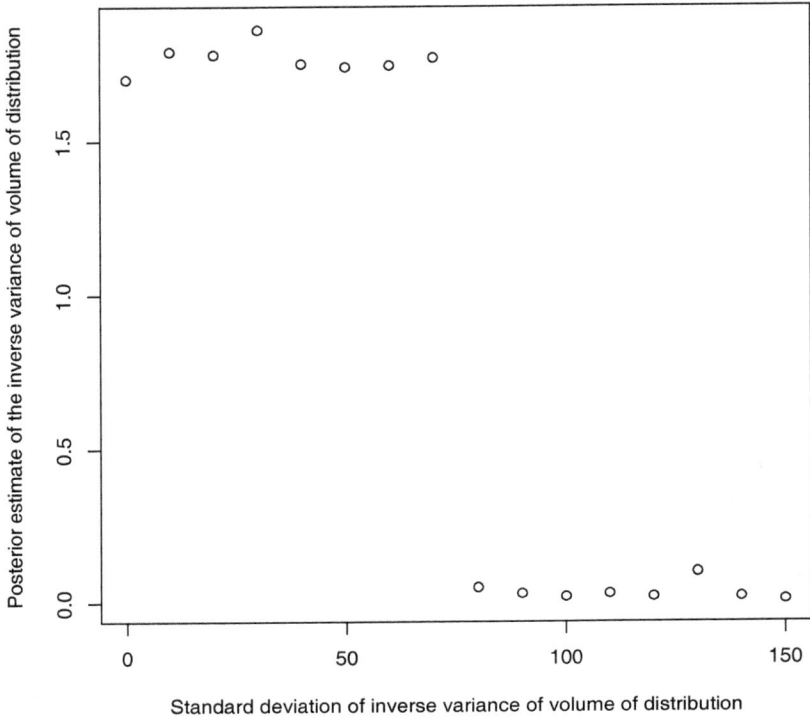

Fig. 5.22.   One-compartment PK model, Case 3. Selection of parameters for $\sigma^{-2}{}_v$.

variance of volume of distribution against the standard deviation of inverse variance of volume of distribution (Figure 5.22). From this figure we can see that the transition from proper and improper posterior distributions occurred at approximately $\sqrt{\xi_v}/\beta_v = 70$. We have selected $\xi_v = 2.5 \times 10^{-5}$, $\beta_v = 10^{-4}$, or $\sqrt{\xi_v}/\beta_v = 50$, located at the boundary of "properness". Therefore, the hyper–parameters of the model $(\lambda, \sigma_0, V_0, \sigma_0^v, \sigma_e, \xi, \beta, \xi_v, \beta_v)$ were chosen as:

$$\lambda = 0.36, \; \sigma_0 = 1, \; V_0 = 20, \; \sigma_0^v = 10,$$
$$\sigma_e = 0.1, \; \xi = 30.8, \; \beta = 0.1, \tag{5.22}$$
$$\xi_v = 2.5 \times 10^{-5}, \; \beta_v = 10^{-4}.$$

The unknown parameters were $(K, w_k, \mu_k, v_k, \sigma_k, \sigma_k^v)$. We analyzed the model in Eqs. (5.17), (5.21) with parameters in Eq. (5.22) with KLMCMC.

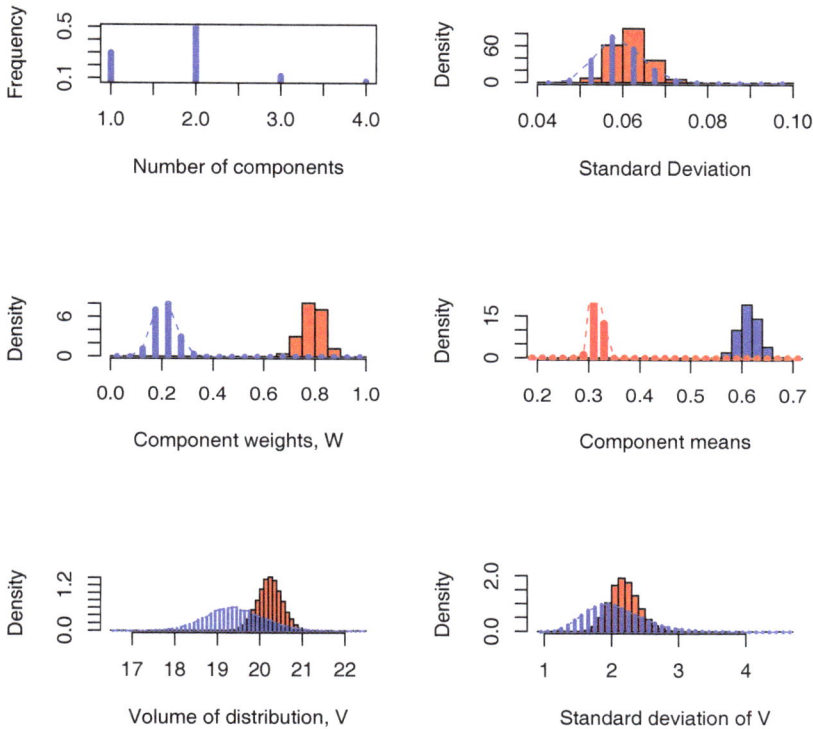

Fig. 5.23. One-compartment PK model, Case 3. Posterior distribution of the number of components (top left), standard deviation σ (top right); Conditional on $K = 2$: component weights (middle left), component means (middle right), volume of distribution (bottom left), standard deviation of the volume of distribution (bottom right).

Table 5.10 One-compartment PK model, Case 3. Results of application of relabeling algorithm to the $K = 2$ model, bimodal V, with randomization.

Iter	$\hat{w}_1$	$\hat{w}_2$	$\hat{\mu}_1$	$\hat{\mu}_2$	$\hat{v}_1$	$\hat{v}_2$	$\hat{\sigma}_1$	$\hat{\sigma}_2$	$\hat{\sigma}_1^v$	$\hat{\sigma}_2^v$
1	0.50	0.50	0.37	0.37	19.86	19.86	0.06	0.06	2.11	2.11
2	0.79	0.21	0.31	0.61	20.25	19.41	0.06	0.06	2.19	2.05

We used 500 birth–death steps of the algorithm (each including 1,000 WinBUGS iterations) with randomization followed by the relabeling step. Posterior distributions are shown in Figure 5.23, output of relabeling is presented in Table 5.10.

Case 4

The next step is to assume that σ_e is unknown. We assume that the measurement precision $\tau_e = \frac{1}{\sigma_e^2}$ has a gamma distribution $G(1, 0.01)$ with mean $E\tau_e = 100$. Therefore, one more equation must be added to the model in Eqs. (5.17) and (5.21):

$$\tau_e = \sigma_e^{-2} \sim G(\cdot | \xi_e, \beta_e), \tag{5.23}$$

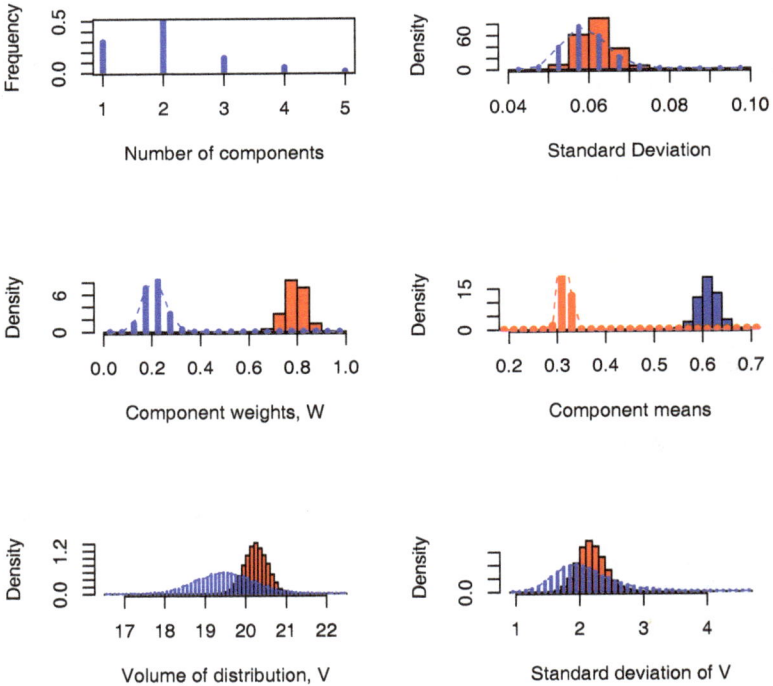

Fig. 5.24.   One-compartment PK model, Case 4. Posterior distribution of the number of components (top left), standard deviation σ (top right); Conditional on $K = 2$: component weights (middle left), component means (middle right), volume of distribution (bottom left), standard deviation of the volume of distribution (bottom right).

Table 5.11   One-compartment PK model, Case 4. Results of application of relabeling algorithm to the $K = 2$ model, bimodal V, unknown σ_e..

Iter	$\hat{w}_1$	$\hat{w}_2$	$\hat{\mu}_1$	$\hat{\mu}_2$	$\hat{v}_1$	$\hat{v}_2$	$\hat{\sigma}_1$	$\hat{\sigma}_2$	$\hat{\sigma}_1^v$	$\hat{\sigma}_2^v$
1	0.50	0.50	0.38	0.38	20.02	20.03	0.06	0.06	2.09	2.09
2	0.79	0.21	0.31	0.61	20.26	19.49	0.06	0.06	2.19	2.06

and the model parameters are now:

$$\lambda = 0.36, \ \sigma_0 = 1, \ V_0 = 20, \ \sigma_0^v = 10,$$

$$\xi = 30, \ \beta = 0.1, \ \xi_e = 1, \ \beta_e = 0.01, \tag{5.24}$$

$$\xi_v = 2.5 \times 10^{-5}, \ \beta_v = 10^{-4}.$$

The unknown parameters were $(K, w_k, \mu_k, v_k, \sigma_k, \sigma_k^v, \sigma_e)$. Table 5.11 and Figure 5.24 show output of the KLMCMC algorithm. The results are essentially the same as Case 3.

Case 5

To test the limits of the KLMCMC algorithm we generate a new, noisier, dataset with a higher "measurement error", $\sigma_e = 0.36$. We generated 500 observations according to the population model in Eqs. (5.17), (5.18), and (5.19) with parameters in Eq. (5.25):

$$\mu_1 = 0.3, \ \mu_2 = 0.6, \ \sigma_{1,2} = 0.06, \ w_1 = 0.8, \ w_2 = 0.2,$$

$$V_0 = 20, \ \sigma_0^v = 2, \ \sigma_e = 0.32, \ \beta_e = 0.1, \tag{5.25}$$

$$\xi_v = 2.5 \times 10^{-5}, \ \beta_v = 10^{-4}, \ \xi_e = 1.$$

Model parameters are equal to

$$\lambda = 0.36, \ \sigma_0 = 1, \ V_0 = 20, \ \sigma_0^v = 10,$$

$$\xi_v = 2.5 \times 10^{-5}, \ \beta_v = 10^{-4}, \tag{5.26}$$

$$\xi = 30.8, \ \beta = 0.1, \ , \xi_e = 1, \ \beta_e = 0.1.$$

The unknown parameters were $(K, w_k, \mu_k, v_k, \sigma_k, \sigma_k^v, \sigma_e)$. We used 500 birth–death steps of the KLMCMC algorithm (each including 1,000 Win-BUGS iterations) with randomization followed by the relabeling step. Results are shown in Figures 5.25, 5.26 and Table 5.13.

To determine the correct number of components in this case, we compared results of KLMCMC, maximum likelihood and the deviance information criterion [Spiegelhalter *et al.* (2002)]. The expression for the likelihood

Before relabeling　　　　　　　　**After relabeling**

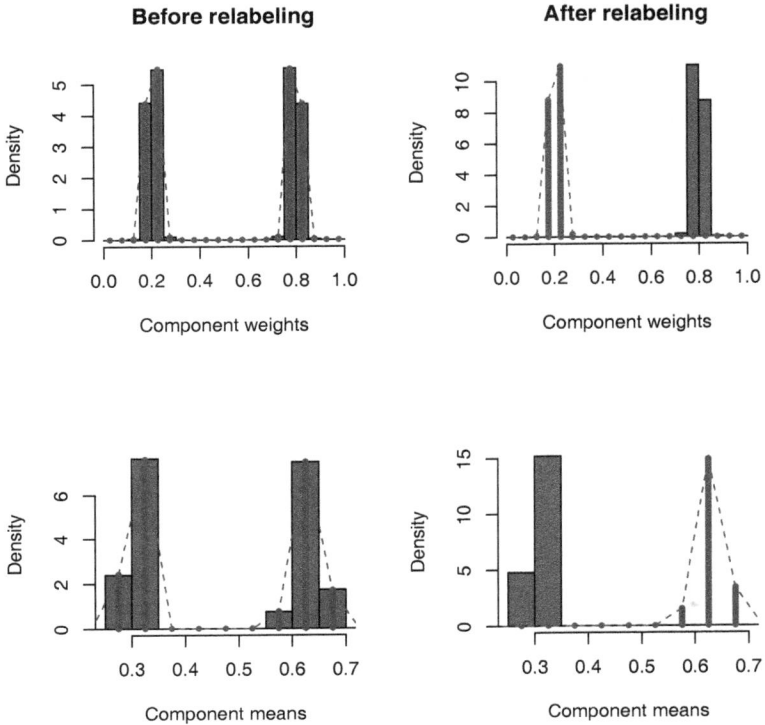

Fig. 5.25.　One-compartment PK model, Case 5: bimodal V, unknown σ_e. Posterior distributions of component weights (top) component means μ (bottom). Right-hand plots correspond to the relabeled distributions.

involves integration of $f(\theta) = \frac{D}{V}e^{-\kappa t}$, over the parameters $\theta = (V, \kappa)$.

$$L(\text{Obs.}|\text{Parms.}) = \prod_{i=1}^{N}\prod_{j=1}^{T} P(Y_{ij}|\text{parameters})$$

$$= \prod_{i=1}^{N}\prod_{j=1}^{T}\sum_{k=1}^{K} w_k \int N(Y_{ij}|f(\theta), \sigma_e^2)N(\theta|\lambda_k, \sigma_k^2)d\theta \qquad (5.27)$$

$$\propto \prod_{i=1}^{N}\prod_{j=1}^{T}\sum_{k=1}^{K} w_k \int e^{-\left(\frac{(Y_{ij}-f(V,\kappa))^2}{2\sigma_e^2} + \frac{(V-v_k)^2}{2\sigma_s^{v2}} + \frac{(\kappa-\mu_k)^2}{2\sigma_k^2}\right)} \frac{dV\,d\kappa}{2\pi\sigma_k^v\sigma_k}.$$

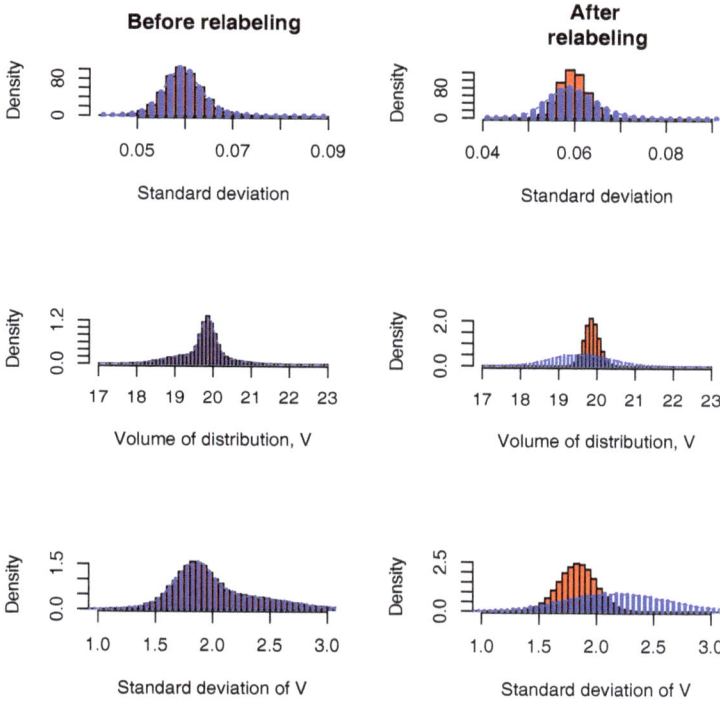

Fig. 5.26. One-compartment PK model, Case 5: bimodal V, unknown σ_e. Posterior distributions of standard deviation σ (top), volume of distribution V (middle), and the standard deviation σ_v of the volume of distribution (bottom). Right-hand plots correspond to the relabeled distributions.

The integrand in Eq. (5.27) is a product of two normal distributions and thus can be expressed as an expected value, namely,

$$\int e^{-\left(\frac{(Y_{ij}-f(\theta))^2}{2\sigma_e^2}+\frac{(\theta-\lambda_k)^2}{2\sigma_k^2}\right)}\frac{d\theta}{2\pi||\sigma_k||^{1/2}} = E_\theta\left(e^{-\left(\frac{(Y_{ij}-f(\theta))^2}{2\sigma_e^2}\right)}\right). \quad (5.28)$$

The likelihood is thus proportional to:

$$L(\text{Obs.}|\text{Parms.}) \propto \prod_{i=1}^N \prod_{j=1}^T \sum_{k=1}^K w_k E_k\left(\frac{e^{-\left(\frac{(Y_{ij}-f(\theta))^2}{2\sigma_e^2}\right)}}{\sqrt{2\pi\sigma_e^2}}\right). \quad (5.29)$$

Integration was performed in MATLAB; results are given in Table 5.12.

Table 5.12   One-compartment PK model, Case 5. Comparison of deviance information, frequency and likelihood for various numbers of components K. Values of deviance information criterion and likelihood are normalized.

Number of components	$K = 1$	$K = 2$	$K = 3$	$K = 4$	$K = 5$
DIC	-17	-24	-21	-18	-18
Frequency	0.3	0.5	0.14	0.05	0.01
Likelihood	4	18	16	15	15

Table 5.13   One-compartment PK model, Case 4. Results of application of relabeling algorithm to the $K = 2$, bimodal V, unknown σ_e, with randomization.

Iter	$\hat{w}_1$	$\hat{w}_2$	$\hat{\mu}_1$	$\hat{\mu}_2$	$\hat{v}_1$	$\hat{v}_2$	$\hat{\sigma}_1$	$\hat{\sigma}_2$	$\hat{\sigma}_1^v$	$\hat{\sigma}_2^v$
1	0.50	0.50	0.42	0.50	19.92	19.92	0.06	0.06	1.96	1.98
2	0.81	0.19	0.30	0.62	19.87	19.98	0.06	0.06	2.02	1.78
3	0.81	0.19	0.30	0.62	19.87	19.98	0.06	0.06	2.02	1.78

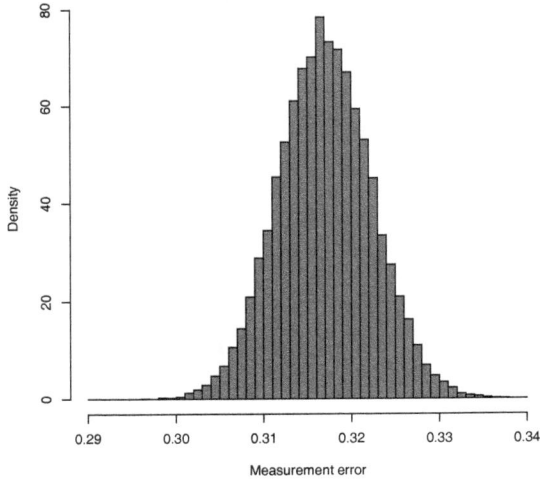

Fig. 5.27.   One-compartment PK model, Case 5. Posterior distribution of σ_e.

The most frequent state of the KLMCMC is $K = 2$, and both the maximum likelihood and the minimum of the deviance information criterion are also attained at $K = 2$ (Table 5.12). Values of model parameters are recovered with at least 5% precision.

We have demonstrated the utility of the KLMCMC method for finding the optimal number of components and model parameters for the one-compartment PK model. Since the method is independent of the type of data used, other types of time-series measurements can be included in the model. We believe that our approach can be used to analyze a patient's specific response to medication. When coupled with genotyping of individuals, it will provide a powerful tool to develop efficient personalized drug treatment that will avoid adverse drug reactions.

5.5 Nonlinear Mixture Models in Gene Expression Studies

Arabidopsis thaliana (Figure 5.28) is an easy-to-grow plant with a well-studied and a relatively short genome (125MB). It has a rapid life cycle – it takes approximately six weeks from germination to mature seed. These factors make it a model for plant genomics, as important as the fruit fly is for animal research. Any living organism adapts to external or internal stimuli by regulating the rate of *transcription* (i.e. synthesizing messenger RNA (mRNA) from the DNA template) of individual genes. To understand and quantify cell processes, researchers measure levels of mRNA in various plant tissues and under different treatment conditions. Such experiments are called *gene expression* experiments. Results of gene expression experiments with *Arabidopsis thaliana* are available to the public through The Arabidopsis Information Resource (TAIR), Stanford Microarray Database, National Center for Biotechnology Information (NCBI), and other online resources.

The AtGenExpress consortium from Germany has conducted a comprehensive analysis of gene expression of *A. thaliana* at various stages of development, uncovering the transcriptome of a multicellular model organism [Weigel *et al.* (2004)]. Analysis of this data helped researchers understand molecular mechanisms of seed development in *A. thaliana*, and can also be extrapolated to other plant species.

For our analysis we have selected a subset of the data (a small part of this subset is represented in Figure 5.29) related to seed development. We want to partition genes into functional classes by their transcription profile and build a mathematical model of transcriptional activation during seed development. There are several dozens genes, called seed-specific genes, which are expressed at higher levels than other genes. They are switched on and off at various stages of seed development according to transcription activation signals.

Fig. 5.28.   A flower of *Arabidopsis thaliana*, courtesy of Dr. Envel Kerdaffrec.

Seed development expression data can be organized as a short time-course of 15 time points: five stages of flowers (whole flowers and carpels), five stages of siliques containing seeds and five stages of seeds only. Within flowers, there are female reproductive organs called carpels, containing ovaries, in which eggs are enclosed, and a sticky structure called the stigma, which serves as receptor to catch pollen (the male reproductive organ containing sperm cells) as shown in Figure 5.30. Fertilized carpels are called siliques. When siliques mature they split open to release seeds.

If we plot expression profiles for differentially expressed genes, we notice that for many of them there are two "regimes": flower and seed. That is

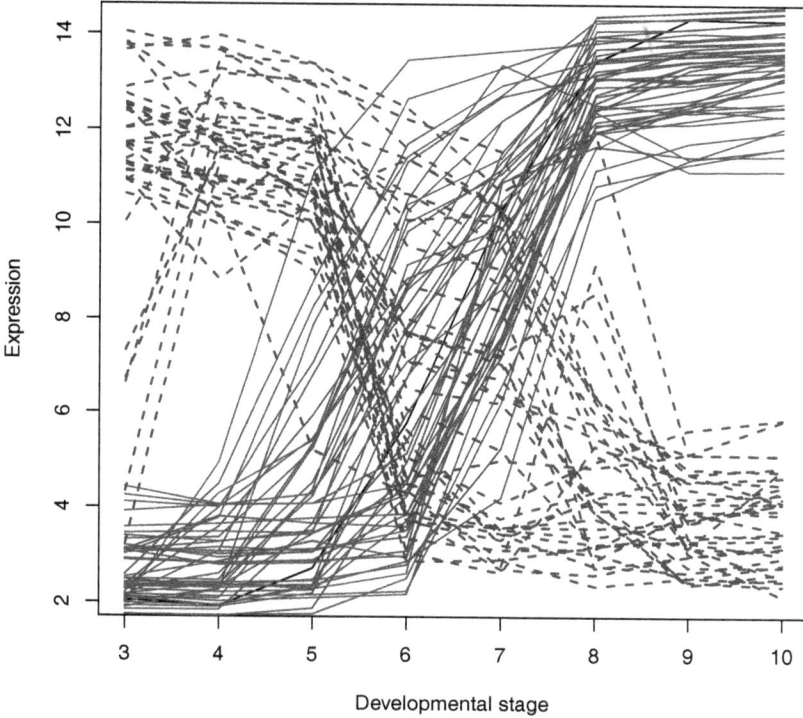

Fig. 5.29. Seed development time series data, temporal profile for 200 most variable genes. X-axis: Developmental stage, Y-axis: Gene expression intensity.

why we model the expression trajectory in the general form as:

$$f(\theta_i, t_j) = \{A_{1,i}t_j + B_{1,i}t_j^2 + C_{1,i}\}e^{-\beta_{1,i}(t_j - G_i)}\eta(G_i - t_j) \tag{5.30}$$
$$+ \{A_{2,i}(t_j - G_i) + B_{2,i}(t_j - G_i)^2 + C_{2,i}\}e^{-\beta_{2,i}(t_j - G_i)}\eta(t_j - G_i),$$

where

$$\eta(x) = \begin{cases} 1 & \text{if } x \geq 0, \\ 0 & \text{otherwise.} \end{cases} \tag{5.31}$$

and $\theta_i = (A_{1,i}, A_{2,i}, B_{1,i}, B_{2,i}, C_{1,i}, C_{2,i}, \beta_{1,i}, \beta_{2,i}, G_i)$ is a nine-dimensional vector of gene-specific parameters. Observations y_{ij} can be described as a sum of trajectory $f(\theta_i, t_j)$ and experimental error term $N(0, \tau^{-1})$, for

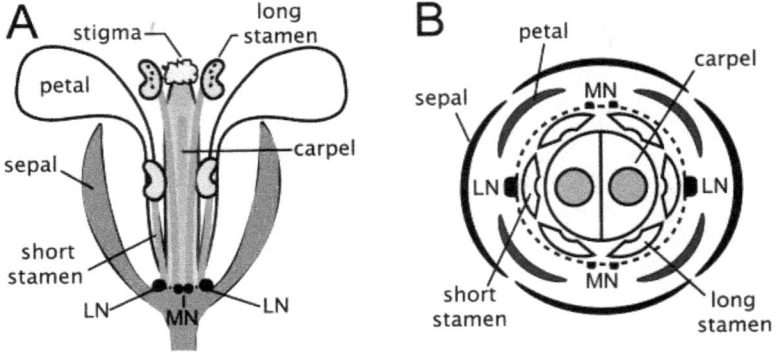

Fig. 5.30. Schematic of *Arabidopsis thaliana* flower and nectarium. (A) Schematic of Arabidopsis flower with front sepal and petals not shown. (B) Schematic cross-section of flower with relative location of floral organs from (A) indicated. The figure is adapted from Kram *et al.* (2009).

simplicity assumed to be common for all genes. We have tested the validity of the model first on the simulated dataset.

5.5.1 *Simulated dataset*

We have applied the KLMCMC method developed in the previous chapter to the simplified model. The seed development model can be expressed as an extension of the one-compartment PK model:

$$y_{ij} \sim N(\lambda_{ij}, \sigma_e^2) \text{ for } i = 1, ..., N \text{ and } j = 1, ..., T, \qquad (5.32)$$

where N is a number of genes, T is a number of experiments, and σ_e is an experimental error. For simplicity we have assumed that there are two distinct groups of genes: "seed-specific" genes and "flower-specific" genes. Thus, the expression of a gene i at time j can be expressed as:

$$\lambda_{ij} = \frac{D}{V_i} e^{(-\mu_i(t_j - G_i))} \eta(t_j - G_i), \qquad (5.33)$$

where G_i is a gene-specific parameter, indicating the starting time when a particular gene is "switched on". Gene-specific parameters (μ_i, v_i, G_i) are assumed to have a mixture distribution:

$$\mu_i \sim \sum_{k=1}^{K} w_k N(\cdot | M_k, \Sigma_k^{\mu 2}),$$

$$v_i \sim \sum_{k=1}^{K} w_k N(\cdot | V_k, \Sigma_k^{v 2}), \tag{5.34}$$

$$G_i \sim \sum_{k=1}^{K} w_k N(\cdot | G_k, \Sigma_k^{G 2}).$$

The component weights are assumed to have a Dirichlet distribution: $w \sim Dir(\cdot | 1, ..., 1)$. Other parameters are given the following prior

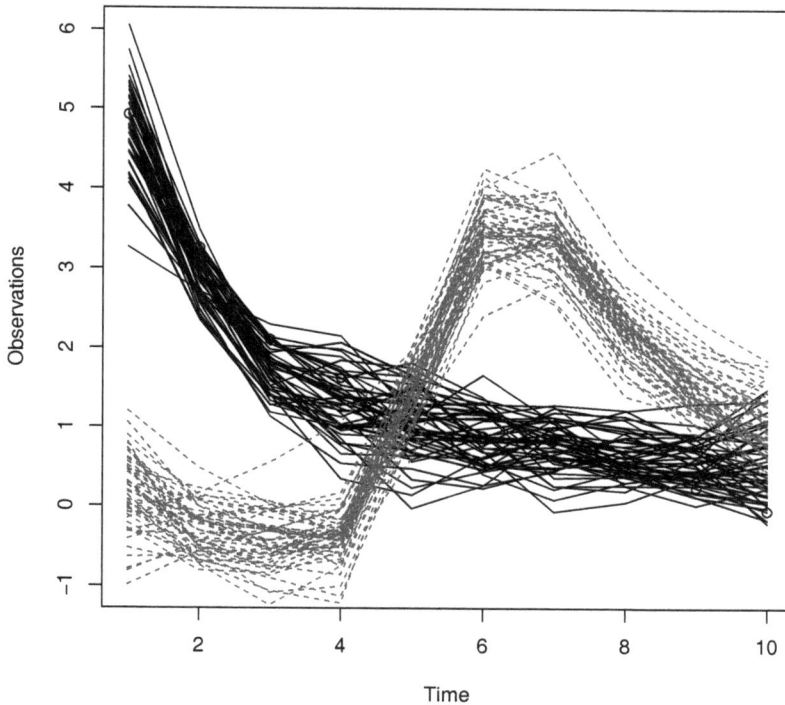

Fig. 5.31. Simulated seed development data.

distributions:

$$G_k \sim U[1, T],$$

$$M_k \sim N(\cdot|M_0, \sigma_m^2), \ V_k \sim N(\cdot|V_0, \sigma_v^2),$$

$$\Sigma_k^{\mu-2} \sim G(\cdot|\xi_m\beta_m), \ \Sigma_k^{v-2} \sim G(\cdot|\xi_v, \beta_v),$$

$$\Sigma_k^{G-2} \sim G(\cdot|\xi_g, \beta_g), \ \sigma_e^{-2} \sim G(\cdot|\xi_e, \beta_e).$$

The model parameters were estimated as:

$$\xi_m = 2.777, \ \beta_m = 0.01,$$

$$\xi_v = 0.025, \ \beta_v = 0.1,$$

$$\xi_g = 1, \ \beta_g = 1,$$

$$\xi_e = 10, \ \beta_e = 0.1,$$

$$M_0 = 0.5, \ V_0 = 20.$$

We have generated measurements for $K = 2$ components, $N = 200$ genes and $T = 10$ time points. We use the following model parameters: $w_{1:2} = \{0.25, 0.75\}$, $G_{1:2} = \{1, 6\}$, $M_{1:2} = \{0.6, 0.3\}$, $V_{1:2} = \{20, 20\}$, $\Sigma_{1:2} = \{0.06, 0.06\}$, $\Sigma_{1:2}^G = \{0.25, 0.25\}$, $\Sigma_{1:2}^v = \{2, 2\}$. The resulting simulated dataset in shown in Figure 5.31.

For this dataset, we ran the KLMCMC algorithm for 500 birth–death steps (each including 1,000 WinBUGS iterations) with randomization followed by the relabeling. We chose $\lambda_b = 2$. In approximately 40% of steps, the chain was in the $K = 2$ state, 25% of steps in the $K = 1$ state, and 20% of steps in the $K = 3$ state, and 15% of steps in the $K = 1$ state. Results are shown in Figures 5.32, 5.33, and Table 5.14.

Table 5.14   Simulated seed example: Results of the 500 iterations of the KLMCMC with RPS.

Iter	w_1	w_2	M_1	M_2	V_1	V_2	G_1	G_2
1	0.5	0.5	0.536	0.536	19.98	19.98	4.72	4.75
2	0.75	0.25	0.609	0.315	20.01	19.89	1	6
3	0.75	0.25	0.609	0.315	20.01	19.89	1	6
4	0.75	0.25	0.609	0.315	20.01	19.89	1	6

Iter	$\Sigma_{\mu,1}$	$\Sigma_{\mu,2}$	$\Sigma_{v,1}$	$\Sigma_{v,2}$	$\Sigma_{G,1}$	$\Sigma_{G,2}$		
1	0.141	0.141	2.074	2.075	2.189	2.174		
2	0.061	0.062	1.975	2.345	0.252	0.146		
3	0.061	0.062	1.975	2.345	0.252	0.146		
4	0.061	0.062	1.975	2.345	0.252	0.146		

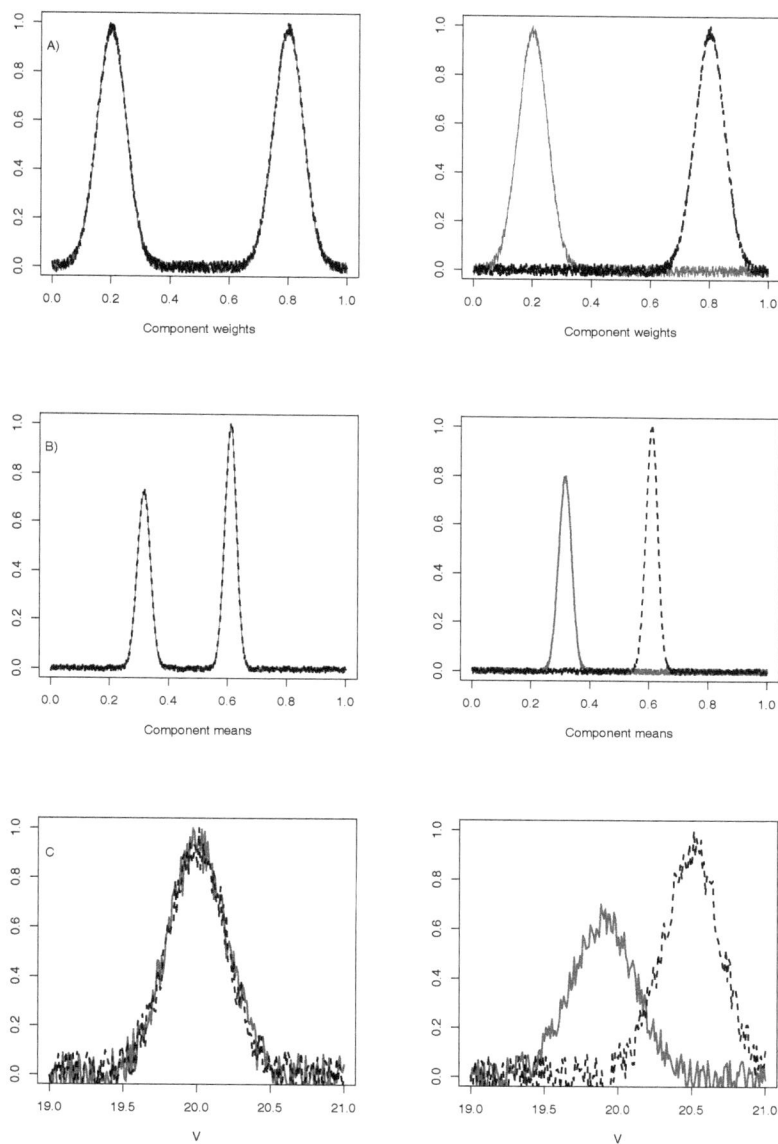

Fig. 5.32. Simulated seed development data before (left) and after (right) relabeling: (A) component weights, (B) M, (C) V.

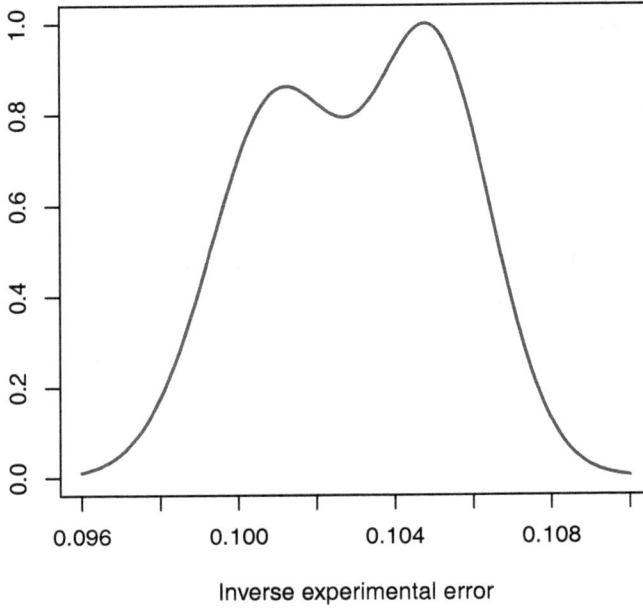

Fig. 5.33.   Simulated seed development data: inverse experimental error $\sigma_e{}^{-2}$.

Chapter 6

Nonparametric Methods

In the previous chapters we have studied parametric mixture models where the mixture components were continuous parameterized densities. In this chapter we consider the nonparametric case where the mixture model is a sum of delta functions. We consider both maximum likelihood and Bayesian cases.

The purpose of this chapter is to explain how the nonparametric maximum likelihood (NPML) and nonparametric Bayesian (NPB) methods are used to estimate an unknown probability distribution from noisy data.

In Section 6.1, we define the basic model used in this chapter. In Section 6.2, we show that the NPML estimation problem can be cast in the form of a primal-dual interior-point method. Then details of an efficient computational algorithm are given. For NPML, nonlinear problems are no harder than linear problems. This is illustrated with the nonlinear pharmacokinetic (PK) model.

In Section 6.3, we define the Dirichlet process for NPB and discuss some of its most important properties. In Section 6.4, we describe the Gibbs sampler for the Dirichlet process and discuss some technical considerations. In Section 6.5, we give a number of NPB examples using the Gibbs sampler, including, among others, the common benchmark datasets: Galaxy, Thumbtacks and Eye Tracking. In Section 6.6, various technical details about condensing the MCMC output are discussed: cleaning, clustering and dealing with multiplicity.

In Section 6.7, we describe the constructive definition of the Dirichlet process given by Sethuraman (1994), called the stick-breaking process. A natural truncation of the stick-breaking process reduces the Dirichlet process to a finite mixture model with a known number of components. Consequently all the algorithms in Chapters 2–4 will apply to this case. In

Section 6.8, examples are then given to compare results from the truncated stick-breaking process to the full Dirichlet process, including the important nonlinear PK model from earlier chapters. In all the above examples, WinBUGS or JAGS programs are provided. In the last Section 6.9 of this chapter, we show an important connection between NPML and NPB.

6.1   Definition of the Basic Model

Consider the following model of Section 3.1:
$$Y_i \sim f_Y(Y_i|\theta_i, \gamma), \quad i = 1, ..., N, \tag{6.1}$$
where the Y_i are independent but not identically distributed random vectors.

As an example, for the PK setting
$$Y_i = h_i(\theta_i) + e_i, \ e_i \sim N(0, \gamma R_i),$$
so that
$$Y_i \sim N(Y_i - h_i(\theta_i), \gamma R_i),$$
where Y_i is the vector of measurements for the ith subject; θ_i is the vector of subject-specific parameters defined on a space Θ; $h_i(\theta_i)$ is a known continuous vector function defining dose inputs, sampling times, and covariates; and the random noise vectors e_i are normal independent noise vectors, with mean 0 and covariance matrix γR_i, where R_i is known and γ is a possibly unknown scale factor.

The $\{\theta_i\}$ are independent and identically distributed random vectors with a common (but unknown) probability distribution F defined on Θ.

The *population analysis problem* is to estimate F based on the data $Y^N = (Y_1, ..., Y_N)$.

Note that the general model described by Eq. (6.1) can be either semiparametric or nonparametric. If in the expression for normal independent noise vectors $e_i \sim N(0, \gamma R_i)$:

- γ is known, the model in Eq. (6.1) is called "nonparametric".
- γ is unknown, the model in Eq. (6.1) is called "semiparametric".

Sometimes a more general semiparametric model is considered:
$$Y_i = h_i(\theta_i, \beta) + e_i, i = 1, ..., N, \tag{6.2}$$
where β is an unknown parameter. A common semiparametric linear model is of the form
$$Y_i = A_i\theta_i + B_i\beta + e_i, \tag{6.3}$$
where A_i and B_i are known matrices and β is unknown.

6.2 Nonparametric Maximum Likelihood

For simplicity in this section, we assume that γ is known. Then the likelihood function for Eq. (6.1) is given by

$$p(Y^N|F) = \prod_{i=1}^{N} \int f_Y(Y_i|\theta_i)dF(\theta_i). \qquad (6.4)$$

The maximum likelihood estimate F^{ML} maximizes $p(Y^N|F)$ over the space of all distributions defined on Θ.

Although this book is about the Bayesian approach, there are two connections between NPML estimation of F and Bayesian ideas. The first connection is that the optimization of Eq.(6.4) with respect to the distribution F can be viewed as an "empirical Bayes" method to estimate the conditional probability of $\theta^N = (\theta_1, ..., \theta_N)$ given the data Y^N, see Koenker and Mizera (2013). The second connection is that the NPML estimate of F can be approximated by the Bayesian stick-breaking method, see Section 6.9.

The theorems of Carathéodory and results of Mallet (1986) and Lindsay (1983) state that, under suitable hypotheses, F^{ML} can be found in the class of discrete distributions with at most N support points. In this case we write

$$F^{ML} = w_1\delta_{\phi_1} + w_1\delta_{\phi_1} + \cdots + w_1\delta_{\phi_1} = \sum_{k=1}^{K} w_k\delta_{\phi_k}, \qquad (6.5)$$

where ϕ_k are the support points of F^{ML} in Θ; $\{w_k\}$ are the weights such that for all $k = 1, ..., K$ $w_k \geq 0$ and $\sum_{k=1}^{K} w_k = 1$; and δ_ϕ represents the delta distribution on Θ with the defining property $\int g(\theta)d\delta_\phi(\theta) = g(\phi)$ for any continuous function $g(\theta)$. More generally $\int g(\theta)dF^{ML}(\theta) = \sum_{k=1}^{K} w_k g(\phi_k)$. This property of transforming integration into summation is crucial for computational efficiency.

It follows from Eqs.(6.4) and (6.5) that

$$p(Y^N|F) = \sum_{k=1}^{K} w_k p(Y_i|\phi_k). \qquad (6.6)$$

The discrete distribution of Eq. (6.5) reduces the infinite dimensional optimization of Eq. (6.4) over the space of measures to a finite dimensional problem. The finite dimensional optimization of Eq. (6.6) over the weights $w = (w_1, ..., w_K))$ and the support points $\phi = (\phi_1, ..., \phi_K)$ is still formidable. If ϕ_k is q-dimensional, then the optimization of Eq. (6.6) is of dimension

$N \times K \times q \times (K - 1)$ where $K \leq N$. Below we discuss an efficient approach to solving this high-dimensional problem.

In Sections 6.2.1–6.2.3 we will use the following vector-matrix notation: R^n will denote the Euclidean space of n-dimensional column vectors; $w = (w_1, ..., w_K)^T \in R^K$; 1_K and 0_K are the vectors in R^K with components all equal to one and zero, respectively.

6.2.1 *Bender's decomposition*

Let

$$\psi_{i,k} = p(Y_i|\phi_k),$$

$$\Psi(\phi) = [\psi_{i,k}]_{N \times K}, \ \phi = (\phi_1, ..., \phi_K).$$

Assume that the row sums of the matrix $\Psi(\phi)$ are strictly positive. For any K-dimensional vector $z = (z_1, ..., z_K)^T \in R^K$, define the function

$$\Phi(z) = -\sum_{k=1}^{K} log(z_k), \text{ if } z > 0,$$

$$\Phi(z) = +\infty, \text{ otherwise.} \tag{6.7}$$

Maximizing Eq. (6.6) with respect to the weights $w = (w_1, ..., w_K)$ and the support points $\phi = (\phi_1, ..., \phi_K)$ is equivalent to the following optimization problem:

Basic problem *Minimize* $G(\phi, w) = \Phi(\Psi(\phi)w)$ *w.r.t.* (w, ϕ): $\phi \in \Theta^K, w > 0$ *and* $1_K^T w = 1$.

In the above, the abbreviation w.r.t. means "with respect to". The optimization in the basic problem is written in minimization form to be consistent with convex optimization theory. The idea of Bender's decomposition is to define the function $w(\phi)$ that solves the primal (P) sub-problem: For fixed $\phi \in \Theta^K$,

$$P : w(\phi) = argmin\{\Phi(\Psi(\phi)w) \text{ w.r.t. } w : w > 0 \text{ and } 1_K^T w = 1\}. \tag{6.8}$$

The optimization problem in (6.8) is convex in w and can be solved very efficiently. Using this solution $w(\phi)$, the original optimization problem becomes the reduced problem:

Reduced problem. *Minimize* $H(\phi) = \Phi(\Psi(\phi)w(\phi))$ *w.r.t.* ϕ: $\phi \in \Theta^K$, *where* $w(\phi)$ *is the solution to* P.

6.2.2 *Convex optimization over a fixed grid G*

We first focus on the sub-problem given by Eq. (6.8). Let the grid G be a set of the form: $G = \{\phi = (\phi_1, ..., \phi_K) : \phi_i \in \Theta, \ i = 1, ..., K\}$. Assume G is fixed. For simplicity we will suppress the symbol ϕ in $w(\phi)$ and $\Psi(\phi)$.

We now derive the primal-dual problems for Eq. (6.8) and the Karush–Kuhn–Tucker (KKT) conditions. This material is standard in convex optimization theory, so no proofs will be given. An excellent reference is the book by Boyd and Vandenberghe (2004).

First consider the primal problem of Eq. (6.8). The constraint $1_K^T w - 1 = 0$ can be removed by adding a penalty to the objective function. So (P) becomes

$$P' : \text{ Minimize } \Phi(\Psi w) + N(1_K^T w - 1) \text{ w.r.t. } w : w > 0. \tag{6.9}$$

It was shown in Baek (2006) that P' has the same solution as P.

The Lagrangian and dual Lagrangian for the optimization problem P' are

$$L(w, z, y, v) = \Phi(z) + N(1_K^T w - 1) - y^T w + v^T(z - \Psi w), \tag{6.10}$$

$$g(y, v) = \min\{L(w, z, y, v) : \text{ w.r.t } (w, z) \in R^K \times R^N, (w, z) > 0\}.$$

Since $L(w, z, y, v)$ is convex in (w, z), necessary and sufficient conditions for the minimizer of $L(w, z, y, v)$ are the gradient conditions:

$$\frac{\partial}{\partial w} L(w, z, y, v) = N1_N^T - y^T - v^T \Psi, \tag{6.11}$$

$$\frac{\partial}{\partial z} L(w, z, y, v) = -1/z^T - v^T.$$

Regrouping terms in the Lagrangian gives

$$L(w, z, u, v) = \Phi(z) + (N \cdot 1_K^T - y^T - v^T \Psi)w - K + v^T z, \tag{6.12}$$

and substituting the above gradient conditions in Eq.(6.10),

$$g(y, v) = -\Phi(v) - N + v^T z = -\Phi(v), \text{ since } v^T z = N. \tag{6.13}$$

The dual problem is then

$$D : \text{ Minimize } \Phi(v) \text{ w.r.t. } (y, v) : y \geq 0, v \geq 0, y + \Psi^T v = N1_K. \tag{6.14}$$

The KKT conditions then become

$$\begin{aligned} \Psi^T v + y &= N1_K, \\ V\Psi w &= 1_N, \\ Wy &= 0_K, \end{aligned} \tag{6.15}$$

where we have used the two equations $\Psi w = z$ and $v = 1/z$ and where $V = Diag(v)$, $Y = Diag(y)$, $W = Diag(w)$.

The following theorem is derived in Boyd and Vandenberghe (2004).

Theorem 6.1. *Let w^* satisfy the constraints $w^* > 0$ and $1_K^T w^* = 1$. Then w^* solves P and (y^*, v^*) solves D if and only if w^*, y^*, v^* satisfy the KKT conditions* (6.15).

6.2.3   *Interior point methods*

One of the guidelines in convex optimization is to reduce problems with inequality constraints to unconstrained problems. In the primal problem P', there is the constraint $w > 0$. To remove this constraint, the ideal penalty for the objective in P' would be the indicator function:

$$I(w) = 0, \text{ if } w > 0,$$

$$I(w) = +\infty, \text{ if } w \leq 0.$$

(6.16)

But $I(w)$ is not differentiable. So what is done is to add a family of penalties of the form $\mu log(w)$, which approximates $I(w)$ as $\mu \downarrow 0$. This is called a "log barrier" penalty, see Boyd and Vandenberghe (2004). Then the primal problem P' now becomes the family of problems:

$$P_\mu : \text{ Minimize } \Phi(\Psi w) + N(1_K^T w - 1) + \mu log(w) \text{ w.r.t.} w : \mu > 0. \quad (6.17)$$

The solution to P_μ now depends on μ and the KKT conditions become

$$\Psi^T v + y = N1_K,$$
$$V\Psi w = 1_N,$$
$$Wy = \mu 1_K.$$

(6.18)

The change of variables $v = v/N$, $y = y/N$, $w = Nw$ in Eq. (6.18) makes the first condition become $\Psi^T v + y = 1_K$, while the rest remain the same. The interior point method applies Newton's method to the perturbed nonlinear equations, KKT_μ:

$$\Psi^T v + y = 1_K,$$
$$V\Psi w = 1_N,$$
$$Wy = \mu 1_K.$$

(6.19)

where μ is decreased to zero. The procedure terminates when

$$\max[\mu, ||V\Psi w - 1_N||, gap(w,v)] \leq 10^{-12},$$

where

$$gap(w,v) = |\Phi(v) - \Phi(\Psi w)|/(1 + \Phi(\Psi w))$$

is the difference between the current value of primal and dual objectives. Further, $gap(w,v) = 0$ if and only if w solves P and v solves D.

Newton's method. We now give a brief description of Newton's method applied to the nonlinear equations KKT_μ. The vector function $F(w,v,y)$ is defined by

$$F(w,v,y) = [\Psi^T v + y,\ V\Psi w,\ Wy]^T.$$

The gradient of F is then

$$\nabla F(w,v,y) = \begin{bmatrix} 0_{KxK} & \Psi^T & I_K \\ V\Psi & Diag(\Psi w) & 0_{NxK} \\ y & 0_{KxN} & W \end{bmatrix}.$$

The Newton step at a point (w,v,y) is obtained by solving the non-singular linear equations:

$$F(w,v,y) + \nabla F(w,v,y)(\Delta w^T, \Delta v,^T \Delta y^T)^T = [1_K^T,\ 1_N^T,\ \mu \cdot 1_K^T]^T. \quad (6.20)$$

After each solution to Eq. (6.20), the value of μ is reduced at the same rate as the quantity $||W\Psi w - 1_N||$. The Newton step $(w + \Delta w, v + \Delta v, y + \Delta y)$ has to be dampened so that the updated parameters w, v, y are strictly positive. For complete numerical details of the algorithm see Bell (2012).

6.2.4 *Nonparametric adaptive grid algorithm*

The second part of Bender's decomposition is to solve the reduced problem:
Reduced problem. *Minimize $H(\phi) = \Phi(\Psi(\phi)w(\phi))$ w.r.t. ϕ: $\phi \in \Theta^K$, where now $w(\phi)$ is the solution to the convex problem P.*

One method of solving the reduced problem is due to Bradley Bell and James Burke of the University of Washington, see Bell (2012); Baek (2006). In their method the gradient of $H(\phi)$ in the reduced problem is calculated and a Newton-type algorithm is performed to do the required optimization. Care must be taken in this approach as $H(\phi)$ is not necessarily differentiable. To remedy this situation, the reduced problem is also put into a family of optimization problems such as the interior point method of Section 6.2.3 and $H(\phi)$ then depends on the parameter $\mu \downarrow 0$. The new $H_\mu(\phi)$ is differentiable and convergence theorems are proved in Baek (2006).

We solved the reduced problem by an adaptive search over Θ coupled with the primal-dual interior-point method described in Section 6.2.3. This is implemented in the nonparametric adaptive grid (NPAG) algorithm.

NPAG was developed over a number of years at the Laboratory of Applied Pharmacokinetics, University of Southern California (USC) by Schumitzky (1991), Leary *et al.* (2001), and James Burke from the University of Washington [Baek (2006)]. The current user interface to NPAG is presented in Neely *et al.* (2012).

In the USC program NPAG, the solution to the reduced problem is performed by using an adaptive grid $G = \{\phi = (\phi_1, ..., \phi_K) : \phi_i \in \Theta, i = 1, ..., K\}$. Initially a large grid $G = G_0$ is chosen. The table below gives a "rule of thumb" about the size of G_0.

Table 6.1   Size of G_0 as a function of parameter dimension.

Dimension of PK parameter ϕ_i	Number of points in G_0
1	1000
2	2000
3	5000
4	10000
5	20000
6	40000
7	80000

The distribution of points in G_0 should cover the set Θ. Typically, Θ is assumed to be a rectangle R^q. The USC-NPAG program uses the so-called Faure points for the initial grid on Θ [Fox (1986)]. However, a set of uniform random points on Θ does equally well.

Next, the convex problem P is solved on G_0. After this step, many points of G_0 have very small probability, say probability less than 10^{-12}. These low probability points are deleted from G_0 leaving a much smaller new grid G_1. Then the grid G_1 is expanded by adjoining to each point $\phi^1 \in G_1$ the vertices of a hypercube with ϕ^1 as its center. This adds approximately $2^{dim\Theta}$ points to G_1 resulting in an new expanded grid G_2. New points that are outside of Θ are deleted. This cycle is repeated with G_2 replacing G_0. The process is continued until convergence of the corresponding likelihoods is attained. At convergence there are no more points in the last grid than there are subjects, just as the Mallet–Lindsay theorem predicts. An example of this algorithm is shown below. We consider a PK model similar to that of Section 5.4.

In closing this section, we mention that there are other ways of solving the original basic problem that deserve serious attention; see Pilla *et al.* (2006) and Wang (2007).

6.2.5 *Pharmacokinetic population analysis problem*

For NPML, nonlinear problems are no harder than linear problems. This is illustrated with the following nonlinear PK population analysis problem. (For an example of a more realistic clinical PK problem using NPML see Tatarinova *et al.* (2013).)

The example below illustrates the behavior of a population of subjects with 80% being slow eliminators (normal case) and 20% being fast eliminators (outliers). The strength of the NPAG method is that it can discover not only the normal group but also the outliers. This is important for drug regimen design on an individual subject. The model is given by

$$Y_{ij} = \frac{D}{V_i} \exp(-\kappa_i \cdot t_j) + e_{ij}, \ i = 1, .., 20, \ j = 1, ..., 5, \qquad (6.21)$$

where $e_{ij} \sim N(0, 0.25)$, $V_i \sim N(1, 0.04)$, $\kappa_i \sim 0.8 \times N(0.5, 0.0025) + 0.2 \times N(1.5, 0.0225)$, $t_i \in (0.2, 0.4, 0.6, 0.8, 1.0)$, and $D = 20$.

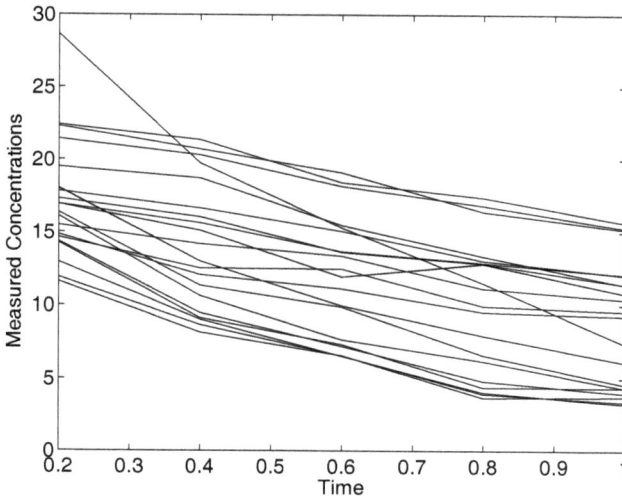

Fig. 6.1.   Graph of trajectories.

A plot of the measured concentrations is given in Figure 6.1. Plots of the "true" density of κ and V are given in Figures 6.2 and 6.3.

The initial grid had 2,129 Fauré points on $\Theta = [0.4, 2.0] \times [0.4, 2.0]$. A plot of these points is shown in Figure 6.4, top, and is compared with

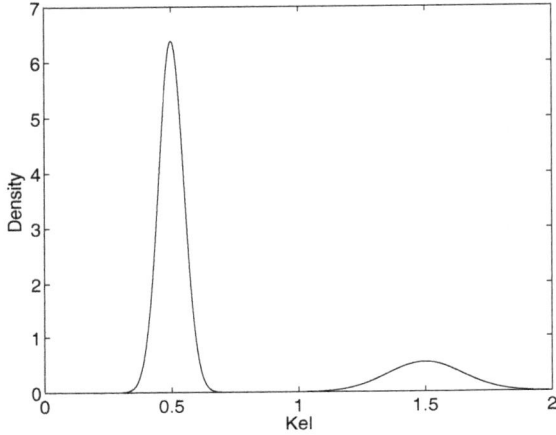

Fig. 6.2.   Density of the elimination constant, κ.

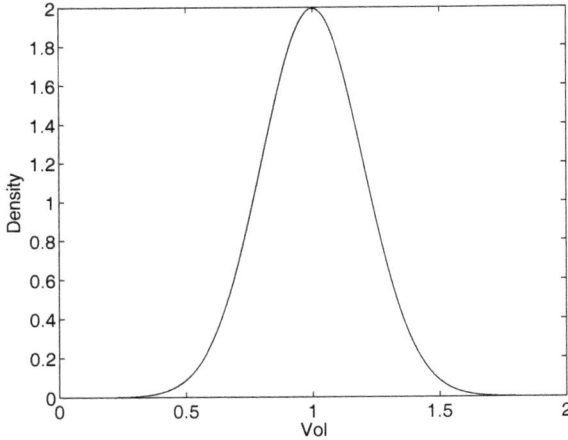

Fig. 6.3.   Density of the volume of distribution, V.

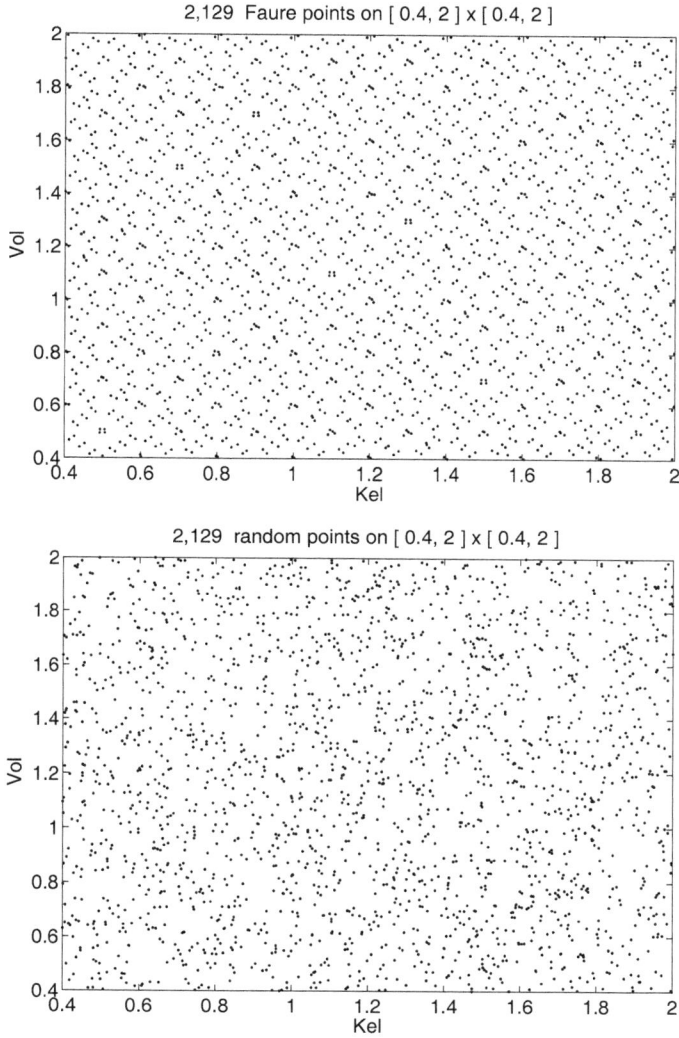

Fig. 6.4. Distribution of 2,129 Faure points (top) and 2,129 random points (bottom) on [0.4, 2.0] × [0.4, 2.0].

a uniform random distribution in Figure 6.4, bottom. The results of the NPAG algorithm are given in Tables 6.2 and 6.3. The discrete marginal distributions for κ and V are shown in Figures 6.5 and 6.6.

Table 6.2   Output of NPAG for the PK model.

Cycle No.	#(Initial Grid)	#(Reduced Grid)	Log-Likelihood
1	2129	18	−110.42
10	90	18	−109.78
20	86	18	−109.13
30	83	17	−108.97
40	83	17	−108.91
50	83	17	−108.897367039
61	83	17	−108.897367039

Table 6.3   Final support points for NPAG method.

κ	V	Probability
0.3916	0.9880	0.0589
0.4450	1.1546	0.0500
0.4498	0.8331	0.1002
0.4649	0.8657	0.0500
0.4811	0.4993	0.0500
0.5015	0.9310	0.0912
0.5054	0.9773	0.0499
0.5173	1.0330	0.0500
0.5299	1.3553	0.0500
0.5313	0.5094	0.0500
0.5325	1.2044	0.0352
0.5329	1.2040	0.0648
0.5584	0.7673	0.0498
1.4410	0.9371	0.0500
1.4862	1.6619	0.0500
1.5650	1.2217	0.1000
1.7378	1.3552	0.0500

6.3   Nonparametric Bayesian Approach

In the nonparametric Bayesian (NPB) approach, the distribution F is itself a random variable. The first requirement then is to place a prior distribution on F. A convenient way of doing this is with the so-called Dirichlet process prior, see Wakefield and Walker (1997, 1998). The main reason we consider the NPB approach is because of its ability to calculate Bayesian credibility intervals no matter what the sample size. This is not possible with other methods, such as the NPML approach.

A Dirichlet process prior on a random distribution F is defined by a probability distribution G_0 and a scale parameter c, where G_0 is our prior guess of F and the parameter c will be the strength of the guess. We write $DP(c, G_0)$ for the Dirichlet process.

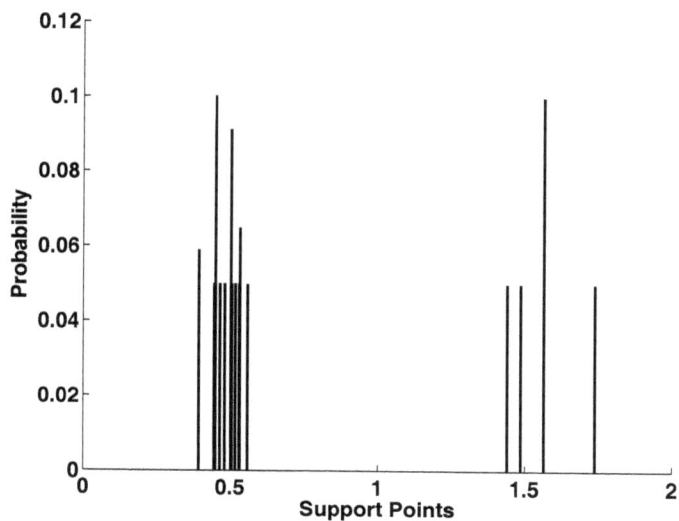

Fig. 6.5.   Marginal density of $F^{ML}(\kappa)$.

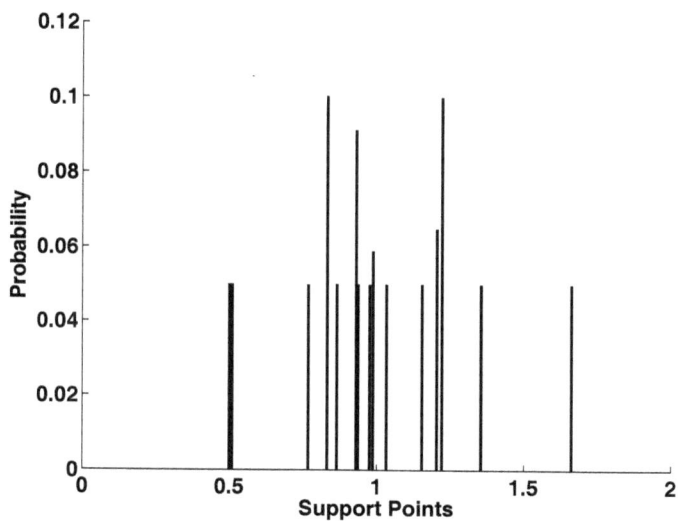

Fig. 6.6.   Marginal density of $F^{ML}(V)$.

In the Bayesian version of the population problem, Eq. (6.1) becomes

$$Y_i \sim f_Y(Y_i|\theta_i), \ i = 1, ..., N,$$

$$\theta_i \sim F, \tag{6.22}$$

$$F \sim DP(c, G_0).$$

There are several important properties of the Dirichlet process (for proofs see Ghosh and Ramamoorthi (2008)). Let $F \sim DP(c, G_0)$ and let $\theta_1, \theta_2, ..., \theta_n$ be n independent samples from F. Then,

$$E[F] = G_0, \tag{6.23}$$

$$E[F|\theta_1, \theta_2, ..., \theta_n] = \frac{1}{c+n} \left(cG_0 + \sum_{i=1}^{n} \delta_{\theta_i} \right), \tag{6.24}$$

$$p(\theta_{n+1}|\theta_1, \theta_2, ..., \theta_n) = \frac{1}{c+n} \left(cG_0 + \sum_{i=1}^{n} \delta_{\theta_i} \right), \tag{6.25}$$

$$p(F|\theta_1, \theta_2, ..., \theta_n) = DP\left(c+n, \frac{1}{c+n} \left(cG_0 + \sum_{i=1}^{n} \delta_{\theta_i} \right) \right). \tag{6.26}$$

From Eqs. (6.24) and (6.25):

$$E[F|\theta_1, \theta_2, ..., \theta_N] = p(\theta_{N+1}|\theta_1, \theta_2, ..., \theta_N). \tag{6.27}$$

Equation (6.25) is very important in the calculation of the Dirichlet process. It can be interpreted as a simple "recipe" for sampling. Sample from $P\{\theta_{n+1}|\theta_1, \theta_2, ..., \theta_n\}$ as follows:

$$\theta_{n+1} = \theta_i, \ \text{w.p.} \ \frac{1}{c+n}, i = 1, ..., n, \tag{6.28}$$

$$\theta_{n+1} \sim G_0, \ \text{w.p.} \ \frac{c}{c+n}, \tag{6.29}$$

where w.p. means "with probability". This last representation implies the "replication" property of the $\{\theta_i\}$. It can be shown that Eq.(6.25) implies that the number of distinct $\{\theta_i\}$ in $\{\theta_1, \theta_2, ..., \theta_n\}$ is order of magnitude $c \times log(n)$ (see Eq. (6.33)). This ultimately limits the computational complexity of the problem.

Not only does the Dirichlet process imply Eq.(6.25), but the converse is also true. This is stated more formally in the next theorem.

Theorem 6.2. *Equivalence theorem*
Let $\theta_1, \theta_2, ..., \theta_n$ be n independent samples from $F \sim DP(c, G_0)$. Then Eq. (6.25) holds. Conversely, if any random sequence $\{\theta_i\}$ satisfies Eq. (6.25), then $\theta_i \sim F$ for some random distribution $F \sim DP(c, G_0)$ [Blackwell and MacQueen (1973)].

Any random sequence satisfying Eq. (6.25) is called a Pólya Urn Process, named after George Pólya, the author of the famous little book *How to Solve It: A New Aspect of Mathematical Method* [Pólya (1957)]. Another interpretation of Eq.(6.25) is given by the Chinese restaurant process [Aldous (1985); Pitman (1996)], describing a Chinese restaurant with an infinite number of infinitely large round tables. The first customer will be seated at an empty table with probability 1. The next customer has two choices: to join the first customer or to pick another empty table. When the $(N+1)^{th}$ customer arrives, she chooses the first unoccupied table with probability $c/(N+c)$ or an occupied i^{th} table with probability $c_i/(N+c)$, where c is the number of people sitting at that table and $c > 0$ is a scalar parameter of the process. Assume that N customers occupy k_N tables. Since

$$\frac{c}{N+c} + \sum_{i}^{k_N} \frac{c_i}{N+c} = \frac{c}{N+c} + \frac{N}{N+c} = 1, \tag{6.30}$$

the above defines a probability distribution. An important property of the Chinese restaurant process is that the expected number of tables occupied by N customers $E[k_N|c]$ grows logarithmically [Gershman and Blei (2012)]. Denote by X_i the Bernoulli random variable such that $X_i = 1$ means that the ith customer chooses to sit alone at a new table, and $X_i = 0$ otherwise. Then,

$$P(X_i = 1) = \frac{c}{i-1+c},$$

$$P(X_i = 0) = 1 - \frac{c}{i-1+c} = \frac{i-1}{i-1+c}. \tag{6.31}$$

The number of occupied tables $k_N = \sum_{i=1}^{N} X_i$, and its expected value is

$$E[k_N|c] = \sum_{i=1}^{N} X_i P(X_i) = \sum_{i=1}^{N} \frac{c}{i-1+c} \times 1 + \sum_{i=1}^{N} \frac{i-1}{i-1+c} \times 0$$

$$= \sum_{i=1}^{N} \frac{c}{i-1+c}. \tag{6.32}$$

Since

$$\int_0^{N-1} \frac{c}{x+c} dx = c(log(N-1+c) - log(c)),$$

$$E[k_N|c] = O(clog(N)), \qquad (6.33)$$

therefore the expected number of tables grows logarithmically with the increase of the number of customers N.

6.3.1   Linear model

Consider the simple model:

$$\begin{aligned} Y_i &= \theta_i, i = 1, ..., N, \\ \theta_i &\sim F, \\ F &\sim DP(1, G_0). \end{aligned} \qquad (6.34)$$

so that we observe directly the parameters $\{\theta_1, \theta_2, ..., \theta_N\}$. Now rewrite Eq. (6.24) as follows:

$$E[F|\theta_1, \theta_2, ..., \theta_N] = \frac{1}{N+1}G_0 + \frac{N}{N+1}\frac{1}{N}\sum_{i=1}^{N}\delta_{\theta_i}. \qquad (6.35)$$

Notice that $\frac{1}{N}\sum_{i=1}^{N}\delta_{\theta_i}$ is the empirical distribution of $\{\theta_1, \theta_2, ..., \theta_N\}$, which is also the NPML estimate F^{ML} of F if $\{\theta_1, \theta_2, ..., \theta_N\}$ are observed. So Eq. (6.24) can be written as

$$E[F|\theta_1, \theta_2, ..., \theta_N] = \frac{1}{N+1}G_0 + \frac{N}{N+1}F^{ML}, \qquad (6.36)$$

as a mixture of the prior estimate and the NPML estimate. For large values of N, the above equation becomes

$$E[F|\theta_1, \theta_2, ..., \theta_N] \approx F^{ML} = \frac{\sum_{i=1}^{N}\delta_{\theta_i}}{N}. \qquad (6.37)$$

Therefore, for large N, data eventually overcome the prior, and the expected posterior of F tends to F^{ML}, the nonparametric maximum likelihood estimate of F.

6.3.2 *Dirichlet distribution*

The Dirichlet process is intimately connected with the Dirichlet distribution. The Dirichlet distribution is a probability distribution on the space of probabilities; while the Dirichlet process is a probability distribution on the space of distributions. Let $Z = (z_1, ..., z_M) \in S^M$ and $\alpha = (\alpha_1, ..., \alpha_M)$, where for all $j = 1, ..., M$ and $\alpha_j > 0$; and where S^M is the set of M-dimensional vectors w, such that $w_m \geq 0$, $\sum_{m=1}^{M} w_m = 1$. Then the random variable Z is said to have a Dirichlet distribution, with parameter α, written $Z \sim Dir(\cdot|\alpha)$, if the density of Z is given by

$$f(Z) = \frac{\Gamma(\sum_{i=1}^{M} \alpha_i)}{\prod_{i=1}^{M} \Gamma(\alpha_i)} z_1^{\alpha_1-1} \times z_2^{\alpha_2-1} \times \cdots \times z_M^{\alpha_M-1}. \qquad (6.38)$$

If $\alpha_1 = 1, ..., \alpha_M = 1$, then Z is uniform on S^M. If $M = 2$, then Z has a beta distribution. In general, for any $j \in 1, ..., M$,

$$E[z_j] = \frac{\alpha_j}{\sum_{m=1}^{M} \alpha_m}. \qquad (6.39)$$

If $(\alpha_1, ..., \alpha_M) \in S^M$ then $\sum_{m=1}^{M} \alpha_m = 1$ so that $E[z_j] = \alpha_j$.

6.3.3 *Dirichlet process for discrete distributions*

The Dirichlet process is a distribution over distributions, i.e., each draw from a Dirichlet process is itself a distribution [Teh (2010)]. If G_0 is a continuous distribution, then the definition of the Dirichlet process requires advanced measure theory [Ferguson (1973)]. If G_0 is a discrete distribution, then the definition of the Dirichlet process only requires basic probability theory.

Theorem 6.3. *Let $F \sim DP(c, G_0))$ and let $\theta_1, ..., \theta_N$ be a sequence of independent samples from G_0. Then the posterior distribution $F|(\theta^N, w^M)$ is a Dirichlet distribution.*

Proof. If $\theta \sim F$, then $\theta = k$ with probability z_k. Let $\theta_1, \theta_2, ..., \theta_N \sim F$. Let n_j denote the number of $\theta_i = q_j$. Let $z^M = (z_1, z_2, ..., z_M)$, $w^M = (w_1, w_2, ..., w_M)$, and $\theta^N = (\theta_1, \theta_2, ..., \theta_N)$. Then by Bayes' law: $p(z^M|\theta^N, w^M) \propto p(\theta^N|z^M, w^M)p(z^M|w^M) \propto \prod_{i=1}^{M} z_i^{n_i} \prod_{i=1}^{M} z_i^{w_i-1} = \prod_{i=1}^{M} z_i^{w_i+n_i-1}$.

Therefore we conclude that $p(z^M|\theta^N, w^M)$ has a Dirichlet distribution $F|\theta^N, w^M \sim p(z^M|\theta^N, w^M) = Dir(\cdot|w^*)$, where $w_j^* = \frac{w_j+n_j}{1+N}$. $\qquad \square$

Let G_0 be a deterministic discrete distribution:

$$G_0 = w_1 \delta_{q_1} + w_2 \delta_{q_2} + \cdots + w_M \delta_{q_M}, (w_1, ..., w_M) \in S^M, \qquad (6.40)$$

Let F be a random distribution such as

$$F = z_1 \delta_{q_1} + z_2 \delta_{q_2} + \cdots + z_M \delta_{q_M}, \text{ where } (z_1, ..., z_M) \sim Dir(\cdot|w_1, ..., w_M) \qquad (6.41)$$

Note that G_0 is deterministic while F is random because the $z_m, m = 1, ..., M$ are random variables. Also note that G_0 is of exactly the same form as the NPML estimate F^{ML} of Section 6.2.

Theorem 6.4. *If*

$$G_0 = \sum_{m=1}^{M} w_m \delta_{q_m}, (w_1, ..., w_M) \in S^M \text{ and}$$

$$F = \sum_{m=1}^{M} z_m \delta_{q_m}, (z_1, ...z_M) \in Dir(\cdot|w_1, ..., w_M)$$

$$then \ F \sim DP(c, G_0), \ where \ c = 1. \qquad (6.42)$$

Remark. The proof of Theorem 6.4 is straightforward but requires more properties of the Dirichlet distribution, see Ranganathan (2006).

Lemma 6.1. *The random distribution defined by Eq. (6.41) also satisfies Eq. (6.25).*

Proof. According to Eq. (6.30), $E[F] = \sum_{m=1}^{M} E[z_{q_m}]\delta_{q_m} = \sum_{m=1}^{M} w_{q_m}\delta_{q_m} = G_0$. Recall that for any events A, B and C $P(A|B)$ can be represented as $\int P(A|C)P(C|B)d\mu_c$. Then,

$$p(\theta_{N+1} = q_j|\theta^N, w^M) = \int p(\theta_{N+1} = q_j|z^M)p(z^M|\theta^N, w^M)dz^M$$

$$= \int z_j p(z^M|\theta^N, w^M)dz^M. \qquad (6.43)$$

Further, $\int z_j p(z^M|\theta^N, w^M)dz^M$ is the mean of z_j with respect to the $Dir(\cdot|w^*)$ distribution, where $w_j^* = \frac{w_j + n_j}{N+1}$. Since $n_j = \sum_{i=1}^{N} \delta_{\theta_i}(q_j)$, then $p(\theta_{N+1} = q_j|\theta^N, w^M) = \frac{1}{N+1}\{w_j + \sum_{i=1}^{N} \delta_{\theta_i}(q_j)\}$. Equivalently, this result can be written as: $p(\theta_{N+1}|\theta^N, w^M) = \frac{1}{N+1}\{G_0(\theta_{N+1}) + \sum_{i=1}^{N} \delta_{\theta_i}(\theta_{N+1})\}$, which gives Property (6.25).   □

Other interesting properties of the Dirichlet processes can be found in Ghosh and Ramamoorthi (2008).

6.4 Gibbs Sampler for the Dirichlet Process

The above results would apply "as is" if the model parameters could be observed exactly as in Eq. (6.34). But in the population model of Eq. (6.22), we only observe the data $Y^N = \{Y_1, Y_2, ..., Y_N\}$ which are noisy nonlinear functions of the $\theta^N = \{\theta_1, \theta_2, ..., \theta_N\}$.

From Eq. (6.25) of the Dirichlet process:

$$
\begin{aligned}
p(\theta_{N+1}|Y^N) &= \int p\left(\theta_{N+1}|\theta^N\right)p(\theta^N|Y^N)\,d\theta^N \\
&= \int \frac{1}{N+1}\left[G_0\left(\theta_{N+1}\right) + \sum_{i=1}^{N} \delta_{\theta_i}(\theta_{N+1})\right]p(\theta^N|Y^N)d\theta^N \\
&= \frac{1}{N+1}G_0(\theta_{N+1})\int p(\theta^N|Y^N)d\theta^N \qquad (6.44) \\
&\quad + \frac{1}{N+1}\sum_{i=1}^{N}\int \delta_{\theta_i}(\theta_{N+1})p(\theta^N|Y^N)d\theta^N \\
&= \frac{1}{N+1}G_0(\theta_{N+1}) + \frac{1}{N+1}\sum_{i=1}^{N}\int \delta_{\theta_i}(\theta_{N+1})p(\theta^N|Y^N)d\theta^N.
\end{aligned}
$$

Suppose we could sample from the posterior probability $p(\theta^N|Y^N)$, obtaining samples $\{(\theta_1^{(m)}, ..., \theta_N^{(m)})\}$, $m = 1, ..., M$. Then the posterior expectation of F is approximately

$$
p(\theta_{N+1}|Y^N) = E(F|Y^N) \approx \frac{1}{N+1}[G_0 + G_M],
$$

$$
G_M = \frac{1}{M}\sum_{m+1}^{M}\sum_{i=1}^{N}\delta_{\theta_i^{(m)}}. \qquad (6.45)
$$

To sample from $p(\theta^N|Y^N)$, we will use the Gibbs sampler, which requires sampling from the full conditional distributions $p(\theta_i|\theta_{-i}, Y_i)$, where $\theta_{-i} = (\theta_1, ..., \theta_{i-1}, \theta_{i+1}, ..., \theta_N)$, for $i = 1, ..., N$. By Bayes' law: $p(\theta_i|\theta_{-i}, Y_i) \propto p(Y_i|\theta_i)p(\theta_i|\theta_{-i})$. The density $p(\theta_i|\theta_{-i})$ can be calculated from Eq. (6.25) by rearranging the order of $\{\theta_1, \theta_2, ..., \theta_N\}$:

$$
p(\theta_i|\theta_{-i}) = \frac{1}{N}\left[G_0(\theta_i) + \sum_{j=1, j\neq i}^{N} \delta_{\theta_j}(\theta_i)\right]. \qquad (6.46)
$$

Therefore, let $(\theta_1^0, ..., \theta_N^0)$ be any point in Θ. For $k = 1, ..., M$ sample θ_i^k, $i = 1, ..., N$ from

$$p(\theta_i^k | \theta_{-i}^k, Y_i^k) \propto p(Y_i | \theta_i^k) G_0(\theta_i^k) + \sum_{j=1, j \neq i}^{N} p(Y_i | \theta_j^k) \delta_{\theta_j^k}(\theta_i^k). \qquad (6.47)$$

This scheme generates a dependent Markov chain $(\theta_1^k, ..., \theta_N^k)$, $k = 1, ..., M$, whose limit is $p(\theta_1, ..., \theta_n | Y^N)$. In the linear case, $p(Y_i | \theta_i) = \theta_i + e_i$, the implementation is straightforward.

6.4.1 *Implementation of the Gibbs sampler*

The Gibbs sampler requires sampling from the full conditional distribution, which could be written as $p(\theta_i | \theta_{-i}, Y^N)$, where $\theta_{-i} = (\theta_1, ..., \theta_{i-1}, \theta_{i+1}, ..., \theta_N)$. We show below that this expression can be written as

$$p(\theta_i | \theta_{-i}, Y^N) = b_i \left[q_{0,i} h(\theta_i | Y_i) + \sum_{j \neq i}^{n} p(Y_i | \theta_j) \delta_{\theta_j}(\theta_i) \right], \qquad (6.48)$$

where

$$b_i^{-1} = q_{0,i} + \sum_{j \neq i}^{n} p(Y_i | \theta_j),$$

$$q_{0,i} = \int G_0(\theta) p(Y_i | \theta) d\theta, \qquad (6.49)$$

$$h(\theta_i | Y_i) = \frac{G_0(\theta) p(Y_i | \theta_i)}{q_{0,i}}.$$

Note that $h(\theta_i | Y_i) = p(\theta_i | Y_i, G_0)$ is the posterior distribution of θ_i given Y_i and assumed prior G_0. Sampling from Eq. (6.48) is performed by

$$\theta_i \neq \theta_j \text{ w.p. } b_i p(Y_i | \theta_j) \text{ for } i = 1, ..., j-1, j+1, ..., n,$$

$$(6.50)$$

$$\theta_i \sim h(\theta_i | Y_i) \text{ w.p. } b_i q_{0,i} \text{ for } i = j.$$

Lemma 6.2. *Equations* (6.48)–(6.50) *define the Gibbs sampler.*

Proof. By Bayes' law: $p(A|B) = \frac{P(B|A)p(A)}{p(B)}$, and $p(A|B, C) = \frac{p(A,B,C)}{p(B,C)} = \frac{p(B|A,C)p(A,C)}{p(B|C)p(C)} = \frac{p(B|A,C)p(A|C)}{p(B|C)}$. Let $A = \theta_i$, $B = Y^N$, $C = \theta_{-i}$, then

$$p(\theta_i | \theta_{-i}, Y^N) = \frac{p(Y^N | \theta_i, \theta_{-i}) p(\theta_i | \theta_{-i})}{p(Y^N | \theta_{-i})}. \qquad (6.51)$$

By the independence of observations,

$$p(Y^N|\theta^N) = \prod_{j=1}^{N} p(Y_j|\theta_j) = p(Y_i|\theta_i) \prod_{j \neq i}^{N} p(Y_j|\theta_j) = C_{1,i} p(Y_i|\theta_i), \quad (6.52)$$

where $C_{1,i} = \prod_{j \neq i}^{N} p(Y_j|\theta_j)$ is a constant independent of θ_i. Then, from Eq. (6.51), $p(\theta_i|\theta_{-i}, Y^N) = C_{1,i} C_{2,i} p(Y_i|\theta_i) p(\theta_i|\theta_{-i})$, where $C_{2,i}^{-1} = p(Y^N|\theta_{-i})$ is a constant independent of θ_i. Note that Eq. (6.27) $(p(\theta_i|\theta_{-i}) = \frac{1}{n+1}[G_0(\theta_i) + \sum_{j \neq i}^{n} \delta_{\theta_j}(\theta_i)])$ can be used for any ordering of $\theta_1, \theta_2, ..., \theta_n$, obtaining

$$p(\theta_i|\theta_{-i}, Y^N) = C_{1,i} C_{2,i} p(Y_i|\theta_i) p(\theta_i|\theta_{-i})$$

$$= C_{1,i} C_{2,i} p(Y_i|\theta_i) \frac{1}{n+1} \left[G_0(\theta_i) + \sum_{j \neq i}^{n} \delta_{\theta_j}(\theta_i) \right] \quad (6.53)$$

$$= C_{1,i} C_{2,i} \frac{1}{n+1} \left[G_0(\theta_i) p(Y_i|\theta_i) + \sum_{j \neq i}^{n} p(Y_i|\theta_i) \delta_{\theta_j}(\theta_i) \right].$$

From the definition of the delta function, the un-normalized density $p(Y_i|\theta_i)\delta_{\theta_j}(\theta_i) = p(Y_i|\theta_j)\delta_{\theta_j}(\theta_i)$. Therefore,

$$p(\theta_i|\theta_{-i}, Y^N) = C_{1,i} C_{2,i} \frac{1}{n+1} \left[G_0(\theta_i) p(Y_i|\theta_i) + \sum_{j \neq i}^{n} p(Y_i|\theta_j) \delta_{\theta_j}(\theta_i) \right].$$
$$(6.54)$$

Denote $q_{0,i} = \int G_0(\theta) p(Y_i|\theta) d\theta$ and $h(\theta_i|Y_i) = \frac{G_0(\theta_i) p(Y_i|\theta_i)}{q_{0,i}}$, then $h(\theta_i|Y_i) = p(\theta_i|Y_i, G_0)$ is the posterior distribution of θ_i conditional on Y_i and assumed prior G_0.

$$p(\theta_i|\theta_{-i}, Y^N) = \frac{C_{1,i} C_{2,i}}{n+1} \left[q_{0,i} h(\theta_i|Y_i) + \sum_{j \neq i}^{n} p(Y_i|\theta_j) \delta_{\theta_j}(\theta_i) \right]. \quad (6.55)$$

Next, denote $b_i = \frac{C_{1,i} C_{2,i}}{n+1}$, and Eq. (6.48) finally becomes

$$p(\theta_i|\theta_{-i}, Y^N) = b_i \left[q_{0,i} h(\theta_i|Y_i) + \sum_{j \neq i}^{n} p(Y_i|\theta_j) \delta_{\theta_j}(\theta_i) \right]. \quad (6.56)$$

Since $p(\theta_i|\theta_{-i}, Y^N)$, $h(\theta_i|Y_i)$ and $\delta_{\theta_j}(\theta_i)$ are probability densities in terms of θ_i, which integrates to 1, we have

$$1 = \int p(\theta_i|\theta_{-i}, Y^N)d\theta_i = b_i \left[q_{0,i} \int h(\theta_i|Y_i)d\theta_i + \sum_{j \neq i}^{n} p(Y_i|\theta_j) \int \delta_{\theta_j}(\theta_i)d\theta_i \right],$$

$$b_i \left[q_{0,i} + \sum_{j \neq i}^{n} p(Y_i|\theta_j) \right] = 1,$$

$$\text{and } b_i^{-1} = q_{0,i} + \sum_{j \neq i}^{n} p(Y_i|\theta_j). \tag{6.57}$$

Sampling from $p(\theta_i|\theta_{-i}, Y^N)$ is therefore performed as follows:

$$\theta_i \neq \theta_j \text{ w.p. } b_i p(Y_i|\theta_j), \text{ for } i = 1, ..., j-1, j+1, ..., n,$$

$$\theta_i \sim h(\theta_i|Y_i) \text{ w.p. } q_{0,i}, \text{ for } i = j, \tag{6.58}$$

which gives Eq. (6.50). □

6.5 Nonparametric Bayesian Examples

To use Eq. (6.50) we must be able to calculate the integral $q_{0,i}$ and sample from $h(\theta_i, Y_i)$. This is straightforward if $G_0(\theta)$ is either discrete or $G_0(\theta)$ is a conjugate prior for $p(Y_i|\theta_i)$. If both $p(Y_i|\theta_i)$ and $G_0(\theta)$ are normally distributed, $G_0(\theta)$ is a conjugate prior for $p(Y_i|\theta_i)$. We now consider a number of examples with conjugate priors.

6.5.1 *Binomial/beta model*

In this section we consider the binomial/beta model. The general model is given by

$$Y_i \sim p_i(Y_i|\theta_i), \ i = 1, ..., N,$$

$$\theta_i \sim F,$$

$$F \sim DP(c, G_0). \tag{6.59}$$

Assume the density of Y_i is binomial based on n_i independent trials and where θ_i is the probability of success. In this case the natural (conjugate) distribution for θ_i is the beta distribution, $Beta(\cdot|a,b)$ with hyper-parameters (a,b) (see Appendix B.1). The model then becomes

$$p(Y_i = k|\theta_i) = Bin(k|n_i, \theta_i) = \binom{n_i}{k} \theta_i^k (1-\theta_i)^{n_i-k}, \ k = 0, 1, ..., n_i,$$

$$\theta_i \sim F,$$

$$(6.60)$$

$$F \sim DP(c, G_0),$$

$$G_0(\theta) = Beta(\theta|a, b) = \frac{\Gamma(a+b)}{\Gamma(a)\Gamma(b)} \theta^{a-1}(1-\theta)^{b-1},$$

The Gibbs sampler equations are then

$$p(\theta_i|\theta_{-i}, Y_i) = b_i \left[q_{0,i} h(\theta_i|Y_i) + \sum_{j \neq i}^{n} p(Y_i|\theta_j)\delta_{\theta_j}(\theta_i) \right]$$

$$= b_i [q_{0,i} h(\theta_i|Y_i) + \binom{n_i}{Y_i} \sum_{j \neq i}^{n} \theta_j^{Y_i}(1-\theta_j)^{n_i-Y_i}\delta_{\theta_j}(\theta_i)], \quad (6.61)$$

where

$$h(\theta_i|Y_i) = Beta(\theta_i|a + Y_i, b + n_i - Y_i),$$

$$(6.62)$$

$$q_{0,i} = \binom{n_i}{Y_i} B_{a,b}(n_i, Y_i),$$

and

$$B_{a,b}(n, y) = \frac{\Gamma(a+b)}{\Gamma(a)\Gamma(b)} \int \theta^{a+y-1}(1-\theta)^{b+n-y-1}d\theta \qquad (6.63)$$

is the *beta* function. Since the term $\binom{n_i}{Y_i}$ cancels we can write

$$p(\theta_i|\theta_{-i}, Y_i) = \tilde{b}_i \left[B_{a,b}(n_i, Y_i)Beta(\theta_i|a + Y_i, b + n_i - Y_i) \right]$$

$$+ \tilde{b}_i \left[\sum_{j \neq i}^{n} \theta_j^{Y_i}(1-\theta_j)^{n_i-Y_i}\delta_{\theta_j}(\theta_i) \right], \qquad (6.64)$$

$$(\tilde{b}_i)^{-1} = B_{a,b}(n_i, Y_i) + \sum_{j \neq i}^{n} \theta_j^{Y_i}(1-\theta_j)^{n_i-Y_i}.$$

Proof of Eqs. (6.61)–(6.64). We have

$$h(\theta_i|Y_i) = \frac{1}{q_{0,i}} G_0(\theta_i)p(Y_i|\theta_i), \text{ where } q_{0,i} = \int G_0(\theta)p(Y_i|\theta)d\theta. \quad (6.65)$$

Substituting for $G_0(\theta)$, we get

$$h(\theta_i|Y_i) = \frac{Beta(\theta_i|a,b)}{q_{0,i}} \binom{n_i}{Y_i} \theta_j^{Y_i}(1-\theta_j)^{n_i-Y_i},$$

$$q_{0,i} = \int Beta(\theta|a,b) \binom{n_i}{Y_i} \theta^{Y_i}(1-\theta)^{n_i-Y_i} d\theta.$$

(6.66)

It follows:

$$h(\theta_i|Y_i) = \frac{Beta(\theta_i|a,b) \binom{n_i}{Y_i} \theta_j^{Y_i}(1-\theta_j)^{n_i-Y_i}}{\int Beta(\theta|a,b) \binom{n_i}{Y_i} \theta^{Y_i}(1-\theta)^{n_i-Y_i} d\theta}$$

$$= \frac{Beta(\theta_i|a,b)\theta_j^{Y_i}(1-\theta_j)^{n_i-Y_i}}{\int Beta(\theta|a,b)\theta^{Y_i}(1-\theta)^{n_i-Y_i} d\theta}$$

$$= \frac{\theta_i^{a-1}(1-\theta_i)^{b-1}\theta_i^{Y_i}(1-\theta_i)^{n_i-Y_i}}{\int \theta^{a-1}(1-\theta)^{b-1}\theta^{Y_i}(1-\theta)^{n_i-Y_i} d\theta}$$

$$= \frac{\theta_i^{a+Y_i-1}(1-\theta_i)^{b-1+n_i-Y_i}}{\int \theta^{a+Y_i-1}(1-\theta)^{b-1+n_i-Y_i} d\theta} = Beta(\theta_i|a+Y_i, b+n_i-Y_i).$$

(6.67)

Finally,

$$q_{0,i} = \int_0^1 \frac{\Gamma(a+b)}{\Gamma(a)\Gamma(b)} \theta^{a-1}(1-\theta)^{b-1} \binom{n_i}{Y_i} \theta^{Y_i}(1-\theta)^{n_i-Y_i} d\theta$$

$$= \binom{n_i}{Y_i} \frac{\Gamma(a+b)}{\Gamma(a)\Gamma(b)} \int_0^1 \theta^{a+Y_i-1}(1-\theta)^{b-1+n_i-Y_i} d\theta = \binom{n_i}{Y_i} B_{a,b}(n_i, Y_i).$$

(6.68)

We will return to this problem in Section 6.8.1.

6.5.2 *Normal prior and linear model*

Although normal linear models rarely occur in practical applications, this case is very useful for benchmark problems such as the Galaxy dataset (see Section 5.1 of this book and Postman *et al.* (1986)).

Assume that

$$G_0 = N(\cdot|\mu, \Sigma), \ Y_i = \theta_i + e_i, \ e_i \sim N(0, R). \tag{6.69}$$

Let $\eta(x, \Sigma)$ be the density of $N(0, \Sigma)$ evaluated at the point x so that $\eta(x-\mu, \Sigma)$ is the normal density with mean vector μ and covariance matrix

Σ. Consider a product of two normal distributions:

$$\eta(x - \mu_1, \Sigma_1) \times \eta(x - \mu_2, \Sigma_2) = \frac{1}{(2\pi)^{k/2}|\Sigma_1|^{1/2}} e^{-\frac{1}{2}(x-\mu_1)^T \Sigma_1^{-1}(x-\mu_1)}$$

$$\times \frac{1}{(2\pi)^{k/2}|\Sigma_2|^{1/2}} e^{-\frac{1}{2}(x-\mu_2)^T \Sigma_2^{-1}(x-\mu_2)} = \frac{1}{(2\pi)^k \sqrt{|\Sigma_1||\Sigma_2|}}$$

$$\times e^{-\frac{1}{2}\{(x-\mu_1)^T \Sigma_1^{-1}(x-\mu_1)+(x-\mu_2)^T \Sigma_2^{-1}(x-\mu_2)\}}. \tag{6.70}$$

The expression in the exponent, $(x-\mu_1)^T \Sigma_1^{-1}(x-\mu_1)+(x-\mu_2)^T \Sigma_2^{-1}(x-\mu_2)$, can be expanded as

$$x^T \Sigma_1^{-1} x + x^T \Sigma_2^{-1} x + \mu_1^T \Sigma_1^{-1} \mu_1 + \mu_2^T \Sigma_2^{-1} \mu_2$$

$$-x^T \Sigma_1^{-1} \mu_1 - \mu_1^T \Sigma_1^{-1} x - x^T \Sigma_2^{-1} \mu_2 - \mu_2^T \Sigma_2^{-1} x.$$

The first two terms can be re-grouped as $x^T(\Sigma_1^{-1}+\Sigma_2^{-1})x$. Denote $\Sigma^* = \{\Sigma_1^{-1} + \Sigma_2^{-1}\}^{-1}$ and let $\mu^* = \Sigma^*\{\Sigma_1^{-1}\mu_1 + \Sigma_2^{-1}\mu_2\}$. Then the expression in the exponent is $(x - \mu^*)^T \Sigma^{*-1}(x - \mu^*)+(\mu_1 - \mu_2)^T(\Sigma_1 + \Sigma_2)^{-1}(\mu_1 - \mu_2)$, and

$$\eta(x - \mu_1, \Sigma_1) \times \eta(x - \mu_2, \Sigma_2) = \frac{1}{(2\pi)^k \sqrt{|\Sigma_1||\Sigma_2|}}$$

$$\times e^{-\frac{1}{2}(\mu_1-\mu_2)^T(\Sigma_1+\Sigma_2)^{-1}(\mu_1-\mu_2)} \times e^{-\frac{1}{2}(x-\mu^*)^T \Sigma^{*-1}(x-\mu^*)}$$

$$= \frac{(2\pi)^{k/2}|\Sigma_1 + \Sigma_2|^{1/2}(2\pi)^{k/2}|\Sigma^*|^{1/2}}{(2\pi)^k \sqrt{|\Sigma_1||\Sigma_2|}} \eta(x - \mu^*, \Sigma^*)\eta(\mu_1 - \mu_2, \Sigma_1 + \Sigma_2). \tag{6.71}$$

Since

$$|\Sigma_1|^{-1/2}|\Sigma_2|^{-1/2} \times |\Sigma_1 + \Sigma_2|^{1/2} \times |\Sigma^*|^{1/2}$$

$$= |\Sigma_1|^{-1/2}|\Sigma_2|^{-1/2} \times |\Sigma_1 + \Sigma_2|^{1/2} \times \frac{|\Sigma_1|^{1/2}|\Sigma_2|^{1/2}}{|\Sigma_1 + \Sigma_2|^{1/2}} \equiv 1,$$

$$\eta(x - \mu_1, \Sigma_1) \times \eta(x - \mu_2, \Sigma_2) = \eta(x - \mu^*, \Sigma^*)\eta(\mu_1 - \mu_2, \Sigma_1 + \Sigma_2),$$

where $\Sigma^{*-1} = \Sigma_1^{-1} + \Sigma_2^{-1}$, and $\mu^* = \Sigma^*(\Sigma_1^{-1}\mu_1 + \Sigma_2^{-1}\mu_2)$,

$$c^* = \eta(\mu_1 - \mu_2, \Sigma_1 + \Sigma_2). \tag{6.72}$$

Note that a product of two normal densities is not a probability density.

6.5.3   *One-dimensional linear case*

For the scalar case, from Eq. (6.72) we obtain

$$\eta(x - \mu_1, \sigma_1^2) \times \eta(x - \mu_2, \sigma_2^2) = \eta(x - \mu^*, \sigma^{*2}) \times \eta(\mu_1 - \mu_2, \sigma_1^2 + \sigma_2^2),$$

$$\text{where } \sigma^{*2} = \frac{\sigma_1^2 \sigma_2^2}{\sigma_1^2 + \sigma_2^2},$$

$$\mu^* = \sigma^{*2} \left(\frac{\mu_1}{\sigma_1^2} + \frac{\mu_2}{\sigma_2^2} \right) = \frac{\sigma_1^2 \sigma_2^2}{\sigma_1^2 + \sigma_2^2} \left(\frac{\mu_1}{\sigma_1^2} + \frac{\mu_2}{\sigma_2^2} \right) = \frac{\mu_1 \sigma_2^2 + \mu_2 \sigma_1^2}{\sigma_1^2 + \sigma_2^2},$$

$$\text{and } \eta(\mu_1 - \mu_2, \sigma_1^2 + \sigma_2^2) = \frac{1}{\sqrt{2\pi(\sigma_1^2 + \sigma_2^2)}} e^{-\frac{(\mu_1 - \mu_2)^2}{2(\sigma_1^2 + \sigma_2^2)}}. \tag{6.73}$$

In the scalar case, the linear model in Eq. (6.69) becomes

$$y_i \sim N(\theta_i, \sigma_1^2),$$
$$G_0 \sim N(\mu, \sigma_2^2). \tag{6.74}$$

Applying Eq. (6.73) we get

$$\eta(y_i - \theta_i, \sigma_1^2) \times \eta(\theta_i - \mu, \sigma_2^2) = \eta(\theta_i - \mu^*, \sigma^{*2})\eta(y_i - \mu, \sigma_1^2 + \sigma_2^2),$$

$$\text{where } \sigma^{*2} = \frac{\sigma_1^2 \sigma_2^2}{\sigma_1^2 + \sigma_2^2}, \text{ and } \mu^* = \frac{y_i \sigma_2^2 + \mu \sigma_1^2}{\sigma_1^2 + \sigma_2^2},$$

$$\text{and } \eta(y_i - \mu, \sigma_1^2 + \sigma_2^2) = \frac{1}{\sqrt{2\pi(\sigma_1^2 + \sigma_2^2)}} e^{-\frac{(y_i - \mu)^2}{2(\sigma_1^2 + \sigma_2^2)}}.$$

In the linear scalar case, Eqs. (6.48)–(6.50) are

$$p(Y_i|\theta_j) = \eta(y_i - \theta_j, \sigma_1^2),$$

$$q_{0,i} = \int G_0(\theta)p(Y_i|\theta)d\theta = \int \eta(\theta - \mu, \sigma_2^2)\eta(y_i - \theta, \sigma_1^2)d\theta$$

$$= c^* \int \eta(\theta - \mu^*, \sigma^{*2})d\theta = c^*, \tag{6.75}$$

$$\text{where } c^* = \frac{1}{\sqrt{2\pi(\sigma_1^2 + \sigma_2^2)}} e^{-\frac{1}{2}\frac{(y_i - \mu)^2}{\sigma_1^2 + \sigma_2^2}},$$

$$\text{and } h(\theta_i|Y_i) = \frac{G_0(\theta_i)p(Y_i|\theta_i)}{q_{0,i}} = c^* \frac{\eta(\theta_i - \mu^*, \sigma^{*2})}{q_{0,i}} = \eta(\theta_i - \mu^*, \sigma^{*2}).$$

To summarize,

$$p(Y_i|\theta_j) = \eta(y_i - \theta_j, \sigma_1^2),$$

$$q_{0,i} = \frac{1}{\sqrt{2\pi(\sigma_1^2 + \sigma_2^2)}} e^{-\frac{1}{2}\frac{(y_i-\mu)^2}{\sigma_1^2+\sigma_2^2}},$$

$$h(\theta_i|Y_i) = \eta(\theta_i - \mu^*, \sigma^{*2}),$$

where $\mu^* = \dfrac{\sigma_2^2 y_i + \sigma_1^2 \mu}{\sigma_1^2 + \sigma_2^2}$ and $\sigma^{*2} = \dfrac{\sigma_1^2 \sigma_2^2}{\sigma_1^2 + \sigma_2^2}.$

(6.76)

6.5.4 *Two-dimensional linear case*

In the two-dimensional case, the model becomes

$$Y_i|\theta_i \sim N(\theta_i, I_2), \ \theta_i \sim G,$$

$$G \sim DP(G_0), \ G_0 \sim N(0, I_2),$$

(6.77)

where $I_2 = \begin{pmatrix} 1 & 0 \\ 0 & 1 \end{pmatrix}$. In Eq. (6.72), $\mu_1 = Y_i$, $\mu_2 = 0$, $\Sigma_1 = \Sigma_2 = I_2$. Then

$$\Sigma^* = (I_2 + I_2)^{-1} = I_2/2,$$

$$\mu^* = \Sigma^*(\Sigma_1^{-1}\mu_1 + \Sigma_2^{-1}\mu_2) = (I_2/2)(Y_i) = Y_i/2,$$

$$c^* = \eta(\mu_1 - \mu_2, \Sigma_1 + \Sigma_2) = \eta(Y_i, 2I_2),$$

(6.78)

$$q_{0,i} = \int G_0(\theta)p(Y_i|\theta)d\theta = \int \eta(\theta - Y_i, I_2)\eta(\theta - 0, I_2)d\theta$$

$$= c^* \int \eta(\theta - \mu^*, \Sigma^{*2})d\theta = c^* = \eta(Y_i, I_2) = \frac{1}{4\pi}e^{-\frac{1}{4}|Y_i|^2}.$$

(6.79)

$$h(\theta_i|Y_i) = \frac{G_0(\theta_i)p(Y_i|\theta_i)}{q_{0,i}} = c^* \frac{\eta(\theta_i - \mu^*, \Sigma^{*2})}{q_{0,i}}$$

$$= \eta\left(\theta_i - \frac{Y_i}{2}, \frac{I_2}{2}\right) \sim N\left(\theta_i|\frac{Y_i}{2}, \frac{I_2}{2}\right).$$

(6.80)

6.5.5 *Plotting the posterior using Gibbs sampling*

Using the notation of Section 6.4, let $F_M = \frac{1}{N+1}[G_0 + G_M]$, where $G_M = \frac{1}{M}\sum_{m=1}^{M}\sum_{i=1}^{N}\delta_{\theta_i^{(m)}}$. Therefore $F_M \approx E(F|Y^N)$. Since

$$\lim_{\sigma \to 0} \frac{1}{\sqrt{2\pi\sigma^2}}exp(-\frac{(x-\mu)^2}{2\sigma^2}) = \delta_\mu(x), \qquad (6.81)$$

the delta function $\delta_{\theta_i^{(m)}}(\theta)$ can be approximated as a normal distribution with mean $\theta_i^{(m)}$ and very small σ_0 (in the next section we will discuss the choice of σ_0). If $\delta_{\theta_i^{(m)}}(\theta) \approx N(\cdot|\theta_i^{(m)}, \sigma_0^2)$, then

$$F_M(\theta) \approx \frac{1}{N+1}\left[G_0(\theta) + \frac{1}{M}\sum_{m=1}^{M}\sum_{i=1}^{N}N(\theta|\theta_i^{(m)}, \sigma_0^2)\right] \qquad (6.82)$$

would be a sum of normals.

6.5.6 *Galaxy dataset*

To illustrate the method, we describe results for the Galaxy dataset problem, which is a benchmark for practically all density estimation programs [Escobar and West (1995); Carlin and Chib (1995); Roeder and Wasserman (1997); Stephens (1997b); Richardson and Green (1997); McLachlan and Peel (2000); Papastamoulis and Iliopoulos (2009)]. It is believed that gravitational forces caused clustering of galaxies. It is possible to measure distances by the red shift in the light spectrum, with the farther galaxies moving at greater velocities [Roeder (1990)]. The dataset consists of 82 galaxy velocities ranging from 9,172 meters per second to 34,479 meters per second, and is described by a linear normal model as in Eq. (6.74).

The model assumes that $Y_i = \theta_i + e_i$, where $e_i \sim N(0,1)$. Parameters of normal distribution $G_0 \sim N(\mu, \sigma_2^2)$ are determined from the mean and variance of observations: $\mu = \overline{Y} = 20.83$ kilometers per second, $\sigma_2 = StDev(Y) = 4.57$ kilometers per second. Using Eq. (6.76), we run the model for 100 iterations, the first 40 iterations are discarded as burn-in.

The posterior distribution F_M of θ is shown in Figure 6.7; the shape of the distribution depends on the value of the "smoothing" parameter σ_0 (Eq. (6.82)). Large values of this parameter (e.g. $\sigma_0 = 2$) force the posterior distribution to resemble the prior $G_0(\theta)$. Small values ($\sigma_0 = 0.1$) result in the nine-modal posterior distribution.

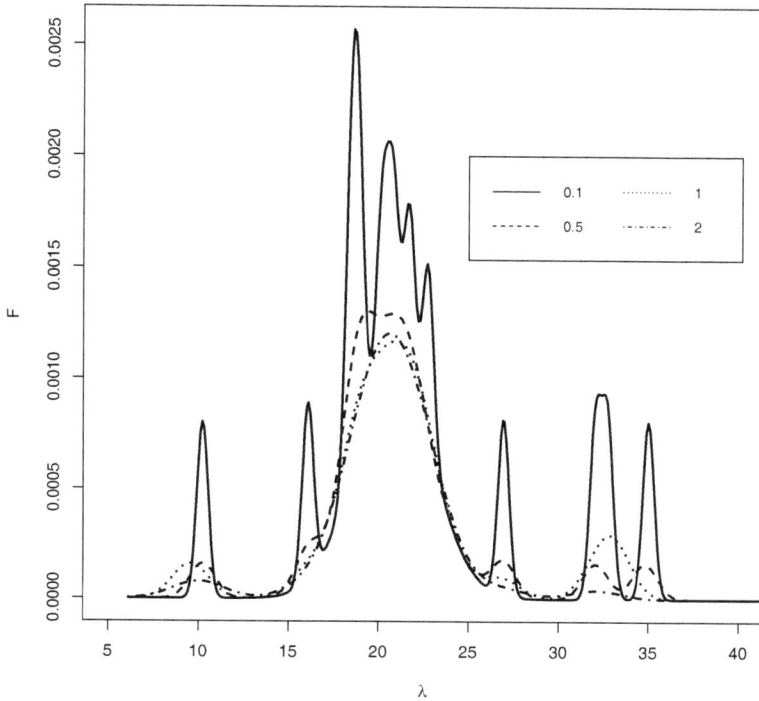

Fig. 6.7. Galaxy dataset. Posterior distribution F_M of θ, measured in kilometers per second, depends on the parameter σ_0 (Eq. (6.82)). Large values of this parameter force posterior distribution to resemble prior $G_0(\theta)$.

6.6 Technical Notes

In order to eliminate rare support points and cluster together support points that are located close to each other in the parameter space, two procedures are implemented: clustering and cleaning.

Cleaning. Consider all support points and for each point calculate the relative frequency with which the Gibbs sampler visits this point. Remove all support points that occur less frequently than some threshold α.

Clustering. The described procedure for clustering is essentially a single linkage hierarchical clustering. Let D be the distance between two clusters. The distance between support points x and y can be computed as a Euclidean distance. At the first step, all pairwise distances between support points are calculated. Initially, every support point is assigned to a separate cluster. Then, iterate:

(1) Find the most similar pair of clusters (i and j) in the current clustering. If the distance $D \leq d_0$, merge clusters (i) and (j). If there are no similar clusters, stop.

(2) Update the distance matrix, D, delete rows and columns corresponding to clusters (i) and (j) and add a row and column corresponding to the new cluster. The distance between the new cluster (i, j) and any old cluster (k) is defined as $d[(k), (i, j)] = min(d[(k), (i)], d[(k), (j)])$.

(3) If all objects are in one cluster, stop. Else, go to step 1.

This procedure produces smaller or equal amounts of support points.

K-means clustering. The K-means method is a logical choice of clustering for the Dirichlet process. The optimal number of clusters K is determined as the mode of the distribution of distinct support points across all runs of the Gibbs sampler. At the first step K random centroids are generated between the smallest and the largest support points. The distance between two points x and y can be computed as a Euclidean distance. Iterate:

(1) For each support point find the nearest centroid.

(2) Assign support points to centroid; for orphan centroids randomly generate new positions.

(3) For each cluster, calculate new positions of the centroid as weighted mean of positions, using support point frequencies as weights.

(4) Repeat Steps 1–3. If clustering allocations do not change, stop.

This procedure produces a smaller or equal amount of the support points.

6.6.1   *Treatment of multiplicity*

The exponential growth of the experimental data in genetic analysis inevitably leads to multiplicity of observation values. Some of the values can be repeated tens and even hundreds of times, while the number of distinct observations may be limited. The brute force approach is not computationally efficient for such datasets. In this section we present a simplified procedure for the treatment of multiplicity.

Let the total number of observations be N as before, however, now there are only L distinct values. Our observations are

$$y_{1,1}, y_{1,2}, ..., y_{1,n_1} \equiv Y_1,$$

$$y_{2,1}, y_{2,2}, ..., y_{2,n_2} \equiv Y_2,$$

$$\vdots$$

$$y_{L,1}, y_{L,2}, ..., y_{L,n_L} \equiv Y_L. \tag{6.83}$$

where $n_i \geq 1$ is the multiplicity of the ith distinct value, such that $\sum_{i=1}^{L} n_i = N$. The support points have the same structure as the data:

$$\theta_{1,1}, \theta_{1,2}, ..., \theta_{1,n_1} \equiv \Theta_1,$$

$$\theta_{2,1}, \theta_{2,2}, ..., \theta_{2,n_2} \equiv \Theta_2,$$

$$\vdots$$

$$\theta_{L,1}, \theta_{L,2}, ..., \theta_{L,n_L} \equiv \Theta_L. \tag{6.84}$$

In this setting, the Gibbs sampler equations, Eq. (6.56)–(6.50), for the normal case with multiplicative observations become

$$p(\Theta_i|\Theta_{-i}, Y^N) = b_i\left[q_{0,i}h(\Theta_i|Y_i) + (n_i - 1)p(Y_i|\Theta_i)\right]$$

$$+ b_i\left[\sum_{j \neq i, j=1,...,L} n_j p(Y_i|\Theta_j)\delta_{\Theta_j}(\Theta_i)\right], \tag{6.85}$$

where

$$b_i^{-1} = q_{0,i} + (n_i - 1)p(Y_i|\Theta_i) + \sum_{j \neq i, j=1,...,L} n_j p(Y_i|\Theta_j). \tag{6.86}$$

Sampling from $p(\Theta_i|\Theta_{-i}, Y^N)$ is performed as follows:

$\Theta_i = \Theta_j$ w.p. $b_i n_j p(Y_i|\Theta_j)$, for $i = 1, ..., j-1, j+1, ..., L-1$,

$\Theta_i = \Theta_i$ w.p. $b_i(n_i - 1)p(Y_i|\Theta_i)$, for $i = j$, $\qquad$ (6.87)

$\Theta_i \sim h(\Theta_i|Y_i)$ w.p. $q_{0,i}$, for $i = L$, where the expressions for $q_{0,i}$ and $p(Y_i|\Theta_j)$ are the same as in the non-multiplicative case, Eq. (6.76):

$$p(Y_i|\Theta_j) = \eta(Y_i - \Theta_j, \sigma_1^2),$$

$$q_{0,i} = \frac{1}{\sqrt{2\pi(\sigma_1^2 + \sigma_2^2)}} e^{-\frac{1}{2}\frac{(Y_i-\mu)^2}{\sigma_1^2+\sigma_2^2}},$$

$$h(\Theta_i|Y_i) = \eta(\Theta_i - \mu^*, \sigma^{*2}), \tag{6.88}$$

$$\text{where } \mu^* = \frac{\sigma_2^2 y_i + \sigma_1^2 \mu}{\sigma_1^2 + \sigma_2^2},$$

$$\text{and } \sigma^{*2} = \frac{\sigma_1^2 \sigma_2^2}{\sigma_1^2 + \sigma_2^2}.$$

It is easy to demonstrate the validity of Eqs. (6.85)–(6.88) from the following considerations. For illustration purposes, use $i = 1$. Since, for a given i, all $\theta_{i,j}$ are the same, the original equation, Eq. (6.56),

$$p(\theta_i|\theta_{-i}, Y^N) = b_i[q_{0,i}h(\theta_i|Y_i) + \sum_{j\neq i}^{n} p(Y_i|\Theta_j)\delta_{\theta_j}(\theta_i)] \tag{6.89}$$

can be re-written using the long-hand notation: $\Theta_1 = \{\theta_{1,1}, \theta_{1,2}, ..., \theta_{1,n_1}\}$ and $Y_1 = \{y_{1,1}, y_{1,2}, ..., y_{1,n_1}\}$, where all $\theta_{1,1} = \theta_{1,2} = ... = \theta_{1,n_1} \equiv \theta_1$ and $y_{1,1} = y_{1,2} = ... = y_{1,n_1} \equiv y_1$ as well. The second term in the expression $\sum_{j\neq 1}^{n} p(Y_1|\Theta_j)\delta_{\theta_j}(\theta_1)$ now corresponds to two different terms with the same value of θ ($\sum_{k=2}^{n_1} p(y_{1,k}|\theta_{1,k})\delta_{\theta_{1,k}}(\theta_{1,k}) = \sum_{k=2}^{n_1} p(y_1|\theta_1)$) and a different ($\sum_{j=2}^{L} \sum_{k=1}^{n_j} p(y_{1,k}|\theta_j)\delta_{\theta_j}(\theta_{1,k})$). Then

$$p(\theta_1|\theta_{-1}, Y^N) = b_1 \left[q_0 h(\theta_1|y_1) + \sum_{k=2}^{n} p(y_1|\theta_1) + \sum_{j=2}^{L}\sum_{k=1}^{n_j} p(y_1|\theta_j)\delta_{\theta_j}(\theta_1) \right]$$

$$= b_1 \left[q_0 h(\theta_1|y_1) + (n_1 - 1)p(y_1|\theta_1) + \sum_{j=2}^{L}\sum_{k=1}^{n_j} n_j p(y_1|\theta_j)\delta_{\theta_j}(\theta_1) \right], \tag{6.90}$$

where n_j is the multiplicity of the jth support point. Extending this result to an arbitrary i, we get

$$p(\theta_i|\theta_{-i}, Y^N) = b_i[q_0 h(\theta_i|y_i) + (n_i - 1)p(y_i|\theta_i) + \sum_{j\neq i} n_j p(y_i|\theta_j)\delta_{\theta_j}(\theta_i)]. \tag{6.91}$$

In the same fashion we treat $b_i^{-1} = q_0 + \sum_{j \neq i} p(y_i|\theta_j)$. Again, set $i = 1$ and use that all $y_{1,k}$ and $\theta_{1,k}$ are the same:

$$b_1^{-1} = q_0 + \sum_{k=2}^{n_1} p(y_{1,k}|\theta_{1,k}) + \sum_{j=2}^{L} \sum_{k=1}^{n_j} p(y_{1,k}|\theta_{1,k})$$

$$= q_0 + (n_1 - 1)p(y_1|\theta_1) + \sum_{j=2}^{L} n_j p(y_1|\theta_1). \tag{6.92}$$

For any i we get

$$b_i^{-1} = q_0 + (n_i - 1)p(y_i|\theta_i) + \sum_{j \neq i}^{L} n_j p(y_i|\theta_j). \tag{6.93}$$

The above multiplicity algorithm can be applied to the treatment of the transcription start site model described in Section 7.5.

6.7 Stick-Breaking Priors

Sethuraman (1994) gave an alternate constructive definition of the Dirichlet process $DP(c, G_0)$. Consider the random distribution defined by

$$F(\cdot) = \sum_{k=1}^{\infty} w_k \delta_{\varphi_k}(\cdot), \tag{6.94}$$

where the random vectors $\{\varphi_k\}$ are iid from a known distribution G_0; δ_φ is the delta distribution which has mass one at φ; and the weights $\{w_k\}$ are now defined by

$$v_k \sim Beta(\cdot|1, \alpha), \ k \geq 1, \ w_1 = v_1.$$

$$w_k = (1 - v_1)(1 - v_2) \cdots (1 - v_{k-1})v_k, k \geq 2, \tag{6.95}$$

where α is either known or has a gamma distribution.

Sethuraman (1994) showed that $\sum_{k=1}^{\infty} w_k = 1$ and that Eq. (6.94) gives a well-defined random distribution. Further, the set of all distributions of the form Eq. (6.94) is identical to $DP(c, G_0)$.

6.7.1 Truncations of the stick-breaking process

Now assume that the sum in Eq. (6.94) is not infinite but finite, i.e. $F(\cdot) = F^K(\cdot) = \sum_{k=1}^{K} w_k \delta_{\varphi_k}(\cdot)$, where K is finite and known. The only change in the model of Eqs. (6.95) is in the weights, where now:

$$v_k \sim Beta(\cdot|1,\alpha), \; k = 1, ..., K-1, \; v_K = 1,$$

$$w_1 = v_1, \; w_k = (1-v_1)(1-v_2)\cdots(1-v_{k-1})v_k, \; 2 \le k \le K. \qquad (6.96)$$

In this case, the basic hierarchical model becomes

$$Y_i \sim f_Y(Y_i|\theta_i,\gamma), \quad i = 1, ..., N,$$

$$\theta_i \sim F^{(K)}(\cdot) = \sum_{k=1}^{K} w_k \delta_{\varphi_k}(\cdot),$$

$$\varphi_k \sim G_0,$$

$$v_k \sim Beta(\cdot|1,\alpha), \; k = 1, ..., K-1; \; v_K = 1, w_1 = v_1,$$

$$w_k = (1-v_1)(1-v_2)\cdots(1-v_{k-1})v_k, \; 2 \le k \le K. \qquad (6.97)$$

This is exactly of the form of the finite mixture model with a known number of components. Consequently, some of the algorithms of Chapters 3 and 4 can now be used without change. An important point to observe is that the above model of Eqs. (6.94) and (6.95) does not have permutation invariance and consequently should not necessarily have label switching. This is because the weights $\{w_k\}$ are not defined by a symmetric Dirichlet distribution but by the stick-breaking definition. In fact, for $k > j$, as observed in Papaspiliopoulos and Roberts (2008) (Section 3.4), there is a non-zero prior probability that $w_k > w_j$, if $|k-j|$ is small. As a result, label switching can occur and the posterior distributions of $w = (w_1, ..., w_K)$ and $\theta = (\theta_1, ..., \theta_K)$ can exhibit multiple modes.

6.7.2 Blocked Gibbs algorithm

The Gibbs sampler for the truncated stick-breaking model is defined in Ishwaran and James (2001). The model in Eq.(6.97) can be written in the following way:

$$Y_i \sim p(Y_i|\phi_{S_i}),$$

$$S_i \sim \sum_{k=1}^{K} w_k \delta_k,$$

$$\phi_k \sim G_0, \qquad\qquad (6.98)$$

$$v_k \sim Beta(\cdot|1, \alpha_0), \; k = 1, ..., K-1, \; v_K = 1,$$

$$w_1 = v_1, \; w_k = (1 - v_1)(1 - v_2) \cdots (1 - v_{k-1})v_k, \; k = 2, ..., K.$$

Since $p(\theta_i = \phi_k) = w_k$, we let $\theta_i = \phi_{S_i}$ where $S_i = k$ with probability w_k.

Note: The set $(S_1, ..., S_N)$ contains values that are not necessarily distinct. For example, let $N = 7$ and $(S_1, ..., S_N) = (1, 2, 5, 7, 2, 7, 2)$. This set of indices $(S_1, ..., S_N)$ corresponds to the set $(\phi_1, \phi_2, \phi_5, \phi_7, \phi_2, \phi_7, \phi_2)$. Then let $(S_1^*, ..., S_m^*)$ be the set of distinct values of $(S_1, ..., S_N)$, in the case of the above example $m = 4$ and $(S_1^*, ..., S_m^*) = (1, 2, 5, 7)$. Note also that the set of distinct values of $(S_1, ..., S_N)$ is not the same as $(1, ..., K)$.

The blocked Gibbs sampler allows sampling from the full posterior distribution of $(\phi^K, S^N, w^K|Y^N)$ by iteratively sampling from the conditional distributions of

$$(\phi^K|S^N, Y^N)$$

$$(S^N|\phi^K, w^K, Y^N)$$

$$(w^K|\phi^K).$$

Then each sample of (ϕ^K, S^N, w^K) will define a random probability measure

$$F^K = \sum_{k=1}^{K} w_k \delta_{\phi_k},$$

which eventually gives the sample from the posterior distribution of F^K.

Note: The posterior distribution of F^K is not the same as the $F|Data$; however, if K goes to infinity, F^K converges a sample from $F|Data$, see Ishwaran and James (2001). The blocked Gibbs sampler requires sampling from the full conditional distributions just like the regular Gibbs sampling.

The following formulas are given in Ishwaran and James (2001):

(1) Sample ϕ_k, $k = 1, .., K$ from: $\phi_k \propto G_0$, if $k \neq \{S_1^*, ..., S_m^*\}$ otherwise

$$\phi_{S_j^*} \sim G_0(\phi_{S_j^*}) \prod_{\{i : S_i = S_j^*\}} p(Y_i | \phi_{S_j^*}), \; j = 1, ..., m.$$

(2) Sample S_i, $i = 1, ..., N$ from

$$p(S_i = s | \phi, w) \sim \sum_{k=1}^{K} w_{k,i} \delta_k(s),$$

$$(w_{1,i}, ..., w_{K,i}) \propto (w_1 p(Y_i | \phi_1), \cdots, w_K p(Y_i | \phi_K)).$$

(3) Sample w_k, $k = 1, ..., K$ from:

$$w_1 = V_1^* \text{ and } w_k = (1 - V_1^*)(1 - V_2^*) \cdots (1 - V_{k-1}^*) V_k^* \; ; \; k = 2, ..., K-1,$$

$$V_k^* = Beta(1 + M_k, \alpha_k + \sum_{i=k+1}^{K} M_i; M_k = \#\{K_i : K_i = k\}).$$

Note: As opposed to the Gibbs sampler for the marginal method, here the Gibbs sampler directly involves the distribution F^K.

6.8   Examples of Stick-Breaking

The problem remains how to truncate the infinite sum in Eq. (6.96). Ishwaran and James (2001) indicated that a relatively small K (e.g. $K < 50$) gives results that are virtually indistinguishable from those based on $DP(c, G_0)$. An elegant way to choose the correct number of truncations is to assume that K is a random variable and use a trans-dimensional method (such as described in Chapter 3 of this book) and to find the joint distribution of (w, θ, K). Given the marginal distribution for the number of components, we choose the value of K_{max} to be at the upper end of the support of this distribution. For this purpose we define a new mixture model representation from the truncated stick-breaking Eqs. (6.96) and (6.97).

$$p(Y_i | w, \phi, K) = \sum_{k=1}^{K} w_k f_Y(Y_i | \phi_k.) \tag{6.99}$$

Note that in this form the likelihood function $p(Y_i | w, \phi, K)$ can be calculated explicitly. We explore this point in examples below.

Table 6.4 Rolling thumbtacks example.

X	0	1	2	3	4	5	6	7	8	9
Count	0	3	13	18	48	47	67	54	51	19

More sophisticated ways of choosing K are based on new results for retrospective sampling [Papaspiliopoulos and Roberts (2008)] and slice sampling [Walker (2007); Kalli *et al.* (2011)]. In these methods, the infinite sum in the stick-breaking representation of Eq. (6.94) is retained, but only as many terms in the sum are used as are needed in the calculation.

6.8.1 *Binomial/beta: Rolling thumbtacks example*

Beckett and Diaconis (1994) generated binary strings from rolls of common thumbtacks. For the dataset, 320 thumbtacks were flicked nine times each. Conditioned on the tack, it is assumed the flicks are independent. Let X denote the number of times each tack landed point up. The data are summarized in Table 6.4.

We employ the binomial/beta model of Section 6.5.1. In this example, $n_i = 9$ and $N = 320$. Since the order of flicks does not matter, the summary of Table 6.4 is sufficient to define the observations $Y_i, i = 1, ..., 320$. Liu (1996) assumed a uniform distribution for G_0 so that $G_0 = Beta(\cdot|1, 1)$ and used a sequential importance sampler (SIS) to determine the approximate posterior mean $E[F|Y]$, for $c = 1$ and $c = 5$. with Liu (1996), and MacEachern and Mueller (1998). For $c = 1$, $E[F|Y]$ was found to be bimodal (Liu, 1996, p. 925) which leads to the interesting conclusion that the two flickers had different styles. For $c = 5$, the bimodality was not as pronounced.

A natural way to analyze this example is to approximate $DP(c = 1, G_0)$ by the truncation of Eqs. (6.94) and (6.95), using $K_{max} = 20$ and implementing the model in JAGS or WinBUGS.

Stick-breaking model for thumbtacks

$$p(Y_i = k|\theta_i) = Bin(k|9, \theta_i), \ i = 0, 1, ..., 320,$$

$$\theta_i \sim F^{(K)}(\cdot) = \sum_{k=1}^{K} w_k \delta_{\varphi_k}(\cdot), \tag{6.100}$$

$$\varphi_k \sim G_0 = Beta(\cdot|1, 1) \equiv U[0, 1],$$

and where $\{w_k\}$ are the stick-breaking weights defined in Eq. (6.96).

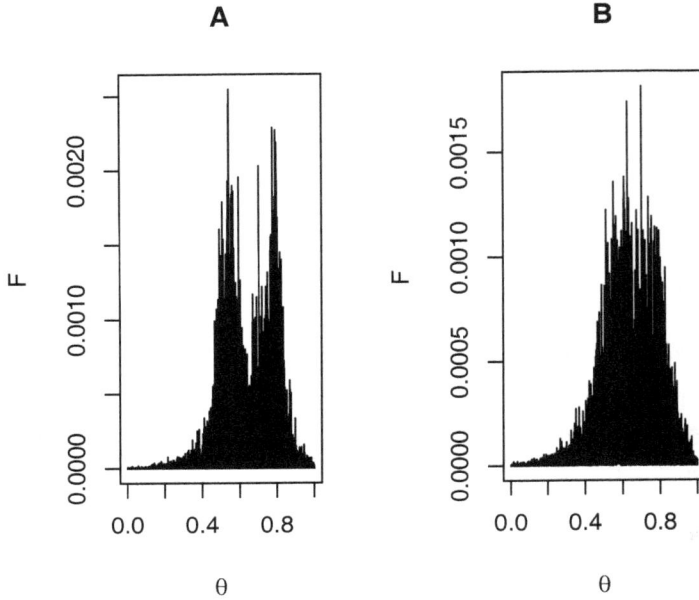

Fig. 6.8.   Binomial/beta model, thumbtack example. 1000 samples from $F|Y$, using two values of parameter c: (A) $c = 1$, (B) $c = 5$.

Following Congdon (2001), we chose the number of stick-breaking components K to be at least 20. WinBUGS was first run for 20,000 iterations with a 1,000 iteration burn-in. The package *coda* was used to get the M samples of w and θ. First, we used these samples to approximate the discrete random distribution $F|Y$:

$$F^{(m)}(\theta) = \sum_{k=1}^{20} w_k^m \delta_{\varphi_k^m}(\theta), \ m = 1, ..., 11. \tag{6.101}$$

In the above equation, w_k^m and $\varphi_k^m, k = 1, ..., 19, m = 1, ..., M$ are the last M samples of the chain. The graph of one of the last $M = 1,000$ samples from $F^{(m)}(\theta)$ for $c = 1$ and $c = 5$ is shown in Figure 6.8. In a similar fashion we obtain samples of the mean value of the distribution $F|Y$, denoted $\mu^{(m)}$:

$$\mu^{(m)} = \int \theta dF^{(m)}(\theta) = \sum_{k=1}^{20} w_k^m \varphi_k^m, \ m = 1, ..., 11. \tag{6.102}$$

The histogram of $\mu^{(m)}, m = 1, ..., 10,000$ is shown in Figure 6.9. In order

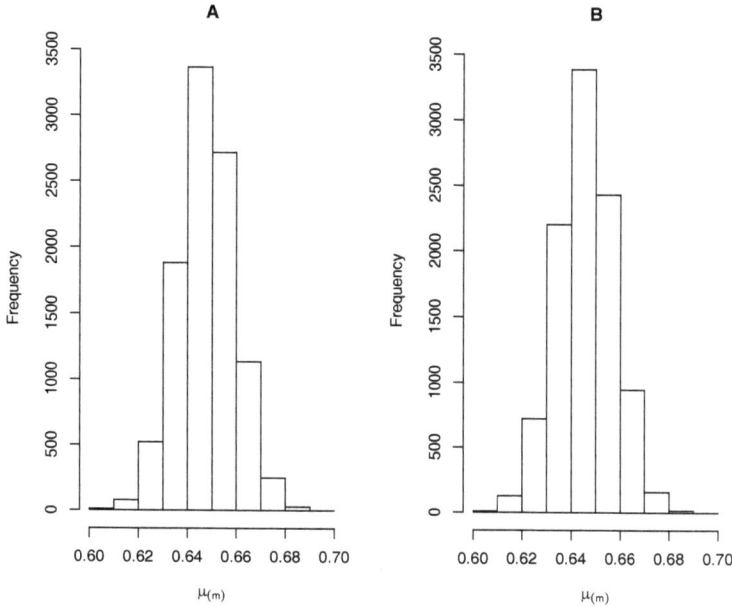

Fig. 6.9. Binomial/beta model, thumbtacks example. Histogram of $\mu^{(m)}, m = 1, ..., 10,000$. (A) $c = 1$ and (B) $c = 5$.

to generate samples of the expected posterior $E[F|Y]$ we calculated

$$E[F|Y] \approx \frac{1}{M} \sum_{m=1}^{M} F^{(m)}(\theta). \qquad (6.103)$$

Figure 6.10 shows the results for $c = 1$ and $c = 5$. Our results agree well with those of Liu (1996) and MacEachern and Mueller (1998). Figure 6.11 shows the histogram of K (the number of active components in the 20-component mixture). These results show that the choice of the value of the precision parameter c makes a big difference.

6.8.2 Determination of the number of mixture components in the binomial/beta model

The birth–death Markov chain Monte Carlo (BDMCMC) method, developed by Matthew Stephens [Stephens (1997a,b, 2000a,b)] is an elegant way to determine the distribution of the number of components (as

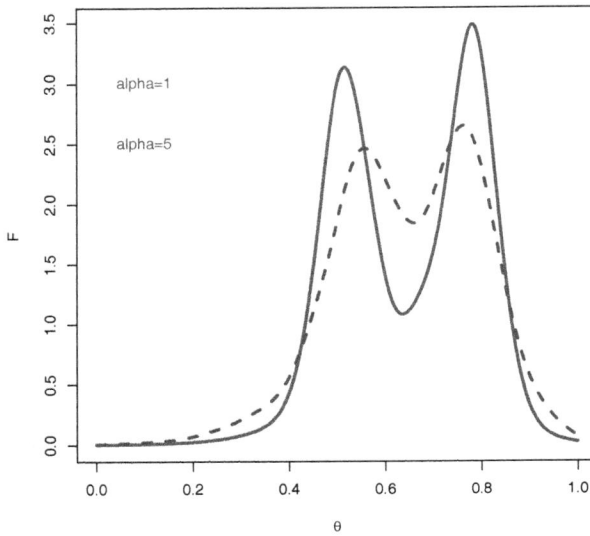

Fig. 6.10.   Binomial/beta model, thumbtacks example. Smoothed version of $E[F|Y]$ for $c = 1$ (solid line) and $c = 5$ (dashed line).

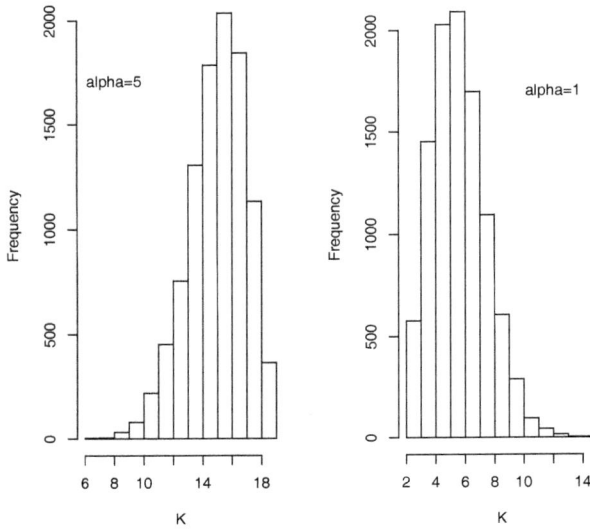

Fig. 6.11.   Binomial/beta model, thumbtacks example. Histogram of K for $\alpha = 1$ and $\alpha = 5$.

described in Section 4.3). For the rolling thumbtacks model of Section 6.8.1 (Eq. (6.100)), the model likelihood is

$$L(Q) = P(Y_1, ..., Y_N | Q, K) = \prod_{i=1,...,N} \sum_{j=1}^{K} w_j Bin(Y_i | n, \varphi_j). \qquad (6.104)$$

$L(Q)$ is used for the selection of a component for the birth or death transdimensional step.

In the birth–death process, the death rate for each component $k \in \{1, ..., K\}$ is calculated as

$$\delta_k(Q) = \lambda_b \frac{L(Q \setminus (w_k, \varphi_k))}{L(Q)} \frac{P(K-1)}{KP(K)} = \frac{L(Q \setminus (w_k, \varphi_k))}{L(Q)} \frac{\lambda_b}{\lambda}$$

$$= \prod_{i=1,...,N} \frac{\sum_{j=1,j \neq k}^{K} w_j Bin(Y_i | n, \varphi_j)}{\sum_{j=1}^{K} w_j Bin(Y_i | n, \varphi_j)} \frac{\lambda_b}{\lambda}. \qquad (6.105)$$

Now, consider another mixture model:

$$Y_i = \theta_i, \ i = 1, ..., N,$$

$$\theta_i \ \sim F^{(K)}(\cdot) = \sum_{k=1}^{K} w_k Bin(\cdot | n, \varphi_k). \qquad (6.106)$$

The death rate $\delta_k(Q)$ for the model in Eq. (6.106) is the same as in Eq. (6.105):

$$\prod_{i=1,...,N} \frac{\sum_{j=1,j \neq k}^{K} w_j Bin(Y_i | n, \varphi_j)}{\sum_{j=1}^{K} w_j Bin(Y_i | n, \varphi_j)} \frac{\lambda_b}{\lambda}.$$

Therefore, in terms of the component death rate computation, the model in Eq. (6.100) is equivalent to the model in Eq. (6.106). Also, the two models result in the identical allocation formulation (see Section 2.2.6) used in WinBUGS computation.

Earlier in this book (Section 4.4, Algorithm 4.2) we developed the KLMCMC algorithm to efficiently determine the number of components in the mixture. KLMCMC uses the weighted Kullback–Leibler (KL) distance in place of the likelihood ratio to determine death rates for individual components. Since it is impossible to define the weighted KL distance for the mixture of delta functions, we suggest using the following trick: using the equivalence of the models in Eq. (6.100) and Eq. (6.106) in terms of likelihood computation, we use the weighted KL distance for Eq. (6.106) to approximate death rates for Eq. (6.100).

In the framework of this approximation, probabilities of trans-dimensional steps (birth P_b or death P_d of mixture components) are determined by the following equations:

$$P_b = \frac{\lambda_B}{\lambda_B + \delta(Q)}, \; P_d = \frac{\delta(Q)}{\lambda_B + \delta(Q)} \; \text{where} \; \delta(Q) = \sum_{k=1}^{K} \frac{1}{\tilde{d}^*(k)}, \quad (6.107)$$

and birth rate λ_B is equal to the prior estimate of the number of components. For the binomial distribution (as shown in Section 4.2.6), the weighted KL distance between the K- and $(K-1)$-component models is

$$d^*(f^{(K)}, f_{k_1,k_2}^{*(K-1)}) = w_{k_1} \left\{ n p_{k_1} log \frac{p_{k_1}}{p^*} + n(1 - p_{k_1}) log \frac{1 - p_{k_1}}{1 - p^*} \right\}$$

$$+ w_{k_2} \left\{ n p_{k_2} log \frac{p_{k_2}}{p^*} + n(1 - p_{k_2}) log \frac{1 - p_{k_2}}{1 - p^*} \right\}, \quad (6.108)$$

where

$$p^* = \frac{w_{k_1} p_{k_1} + w_{k_2} p_{k_2}}{w_{k_1} + w_{k_2}}. \quad (6.109)$$

The distance between a component k and all other components is defined as $\tilde{d}^*(k) = min_{k_1} d^*(f^{(K)}, f_{k,k_1}^{*(K-1)})$ (see Section 4.2.8 for discussion).

Using the equations above, we ran 100 trans-dimensional steps of KLMCMC, each containing 2,000 Gibbs sampler steps, discarding the first 1,000 as burn-in. In this example we used two values of the precision parameter, $c = 1$ and $c = 5$. We used $\lambda_B = 15$. Following Congdon (2001), we chose the maximum possible number of stick-breaking components K_{max} to be 20. As is clear from Figure 6.12 and Table 6.5, for $c = 1$ the number of stick breaks K_{max} can be reduced to 9. This is consistent with the previous analysis by Liu (1996).

When the precision parameter $c = 5$, the largest number of components is $K_{max} = 12$. In order to overcome this ambiguity and dependence on the prior parameter, we considered two scenarios $c \sim G(\cdot|1, 1)$ and $c \sim G(\cdot|0.1, 0.1)$. For the two cases, the distribution of the number of

Table 6.5 Binomial/beta model, thumbtacks example. Number of components as determined by KLMCMC for 100 trans-dimensional runs.

K	3	4	5	6	7	8	9	10	11	12
$c = 1$	1	5	23	35	23	10	3	0	0	0
$c = 5$	1	5	7	9	15	18	17	14	10	4

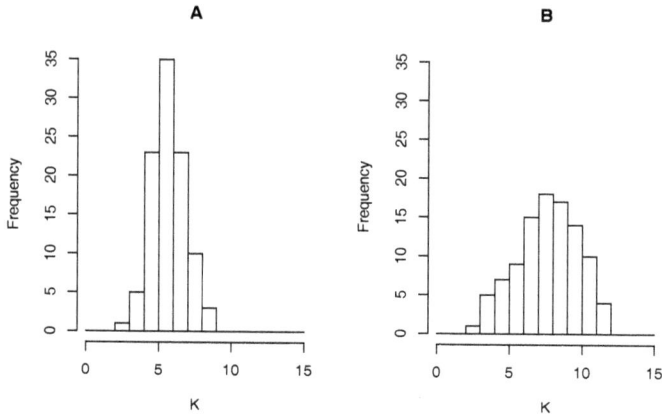

Fig. 6.12. Binomial/beta model, thumbtacks example. Number of components in mixture: (A) $c = 1$ and (B) $c = 5$ for $C = 20$.

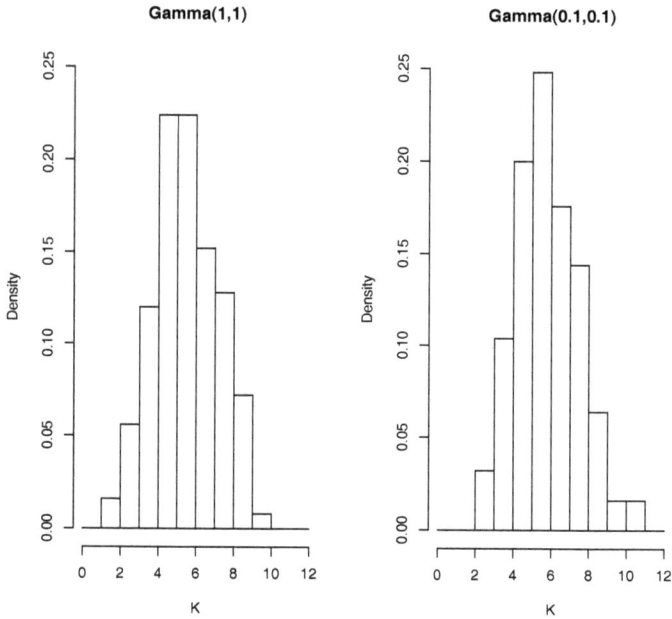

Fig. 6.13. Number of components for 100 trans-dimensional runs of KLMCMC for the binomial/beta example: (A) $c \sim G(\cdot|1, 1)$, (B) $c \sim G(\cdot|0.1, 0.1)$.

mixture components for 100 trans-dimensional runs of KLMCMC for the binomial/beta example is shown in Figure 6.13. In both cases, the mode of the number of mixture components is located at $K = 6$ and the posterior distributions of K are similar. From simultaneous consideration of both distributions we can conclude that $K_{max} = 11$ is most likely sufficient to describe a variety of model settings.

6.8.3   *Poisson/gamma Eye-Tracking example*

The Eye-Tracking example is taken from Congdon (2001, Example 6.27, p. 263). The model is given by Eqs. (6.97) with

$$f_Y(Y_i|\theta_i) = Pois(Y_i|\theta_i),$$
$$\theta_i \sim F^{(K)}(\cdot) = \sum_{k=1}^{K} w_k \delta_{\varphi_k}(\cdot), \qquad (6.110)$$
$$\varphi_k \sim G_0 = G(\cdot|1,1),$$

and where $\{w_k\}$ are the stick-breaking weights defined in Eq. (6.96), with $c = 1$.

There are 101 data points $\{Y_i, i = 1, ..., 101\}$, but there are only 19 distinct values of the set $\{Y_i\}$. The data are presented in Table 6.6. This problem is analyzed using the truncation approach. The corresponding BUGS code (adapted from Congdon (2001)) is given in the appendix. Note that the notations are consistent with Congdon (2001) and specific to this section.

In this case, when $c = 1$ and $G_0(\cdot) = G(\cdot|1,1)$ an interesting phenomenon occurs which allows a good comparison between the Gibbs sampler and truncation methods. Using the Gibbs sampler approach, Escobar and West (1998) noticed that the data observed under the right tail are unusual, and that the posterior $p(\theta|Y_{92})$ (as well as $p(\theta|Y_{70}), ..., p(\theta|Y_{73})$, $p(\theta|Y_{80}), ..., p(\theta|Y_{93})$) is unexpectedly bimodal. The results from BUGS

Table 6.6   Eye-Tracking example. Measurements.

0	0	0	0	0	0	0	0	0	0	0	0	0	0	0	0
0	0	0	0	0	0	0	0	0	0	0	0	0	0	0	0
0	0	0	0	0	0	0	0	0	0	0	0	0	0	1	1
1	1	1	1	1	1	1	1	1	1	1	1	2	2	2	2
2	2	2	2	2	3	3	3	3	4	4	5	5	5	6	6
6	7	7	7	8	9	9	10	10	11	11	12	12	14	15	15
17	17	22	24	34											

Table 6.7 Eye-Tracking example. Results of Gibbs sampler, sample size is 10,000, first 1,000 iterations discarded.

Node	Mean	StDev	MC error	2.5%	Median	97.5%
K	13.8	2.96	0.2018	7.0	14.0	18.0
mu[92]	13.23	3.149	0.04344	5.924	14.05	17.5

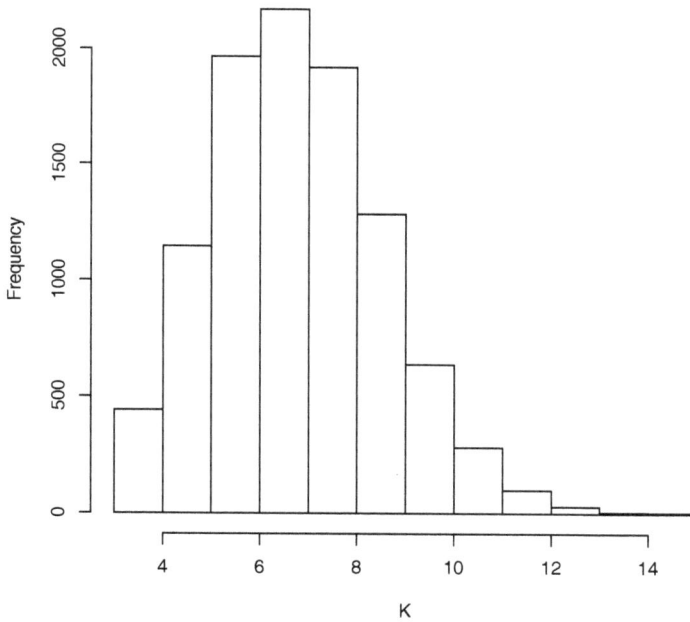

Fig. 6.14. Eye-Tracking example. Histogram of the number of mixture components, K.

(which are similar to Escobar and West (1998)) are shown in Table 6.7 and Figures 6.14 and 6.15.

6.8.4 *Determination of the number of mixture components in Poisson/gamma case*

The number of mixture components was determined using KLMCMC as described in Section 4.4, Algorithm 4.2. Probabilities of trans-dimensional steps (birth P_b or death P_d of mixture components) are determined by the

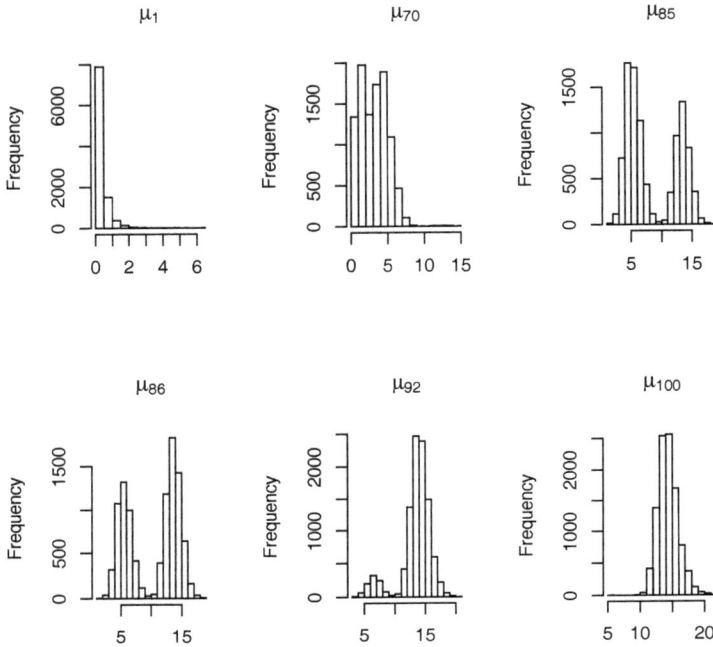

Fig. 6.15. Eye-Tracking example. Posterior distributions $p(\theta|Y_i)$ for $i = 1, 70, 85, 86,$ 92, 100. Graph notation: In terms of Eq. (6.109), $\mu = \theta$.

following equations:

$$P_b = \frac{\lambda_B}{\lambda_B + \delta(Q)}, \quad P_d = \frac{\delta(Q)}{\lambda_B + \delta(Q)}, \text{ where } \delta(Q) = \sum_{k=1}^{K} \frac{1}{\tilde{d}^*(k)}. \quad (6.111)$$

and birth rate λ_B is equal to the prior estimate of the number of components. For the Poisson distribution (as shown in Section 4.2.7), the weighted KL distance between the K- and $(K-1)$-component model is defined in Eq. (6.112):

$$d^*(f^{(K)}, f^{*(K-1)}_{k_1, k_2}) = \sum_{j=k_1, k_2} w_j \lambda_j log \frac{\lambda_j}{\lambda^*}, \text{ where } \lambda^* = \frac{w_{k_1} \lambda_{k_1} + w_{k_2} \lambda_{k_2}}{(w_{k_1} + w_{k_2})}.$$

$$(6.112)$$

Results of the application of KLMCMC to the Eye-Tracking example are presented in Figures 6.16–6.20. A remarkable feature of the Eye-Tracking example is the presence of label switching, evident from the bimodality

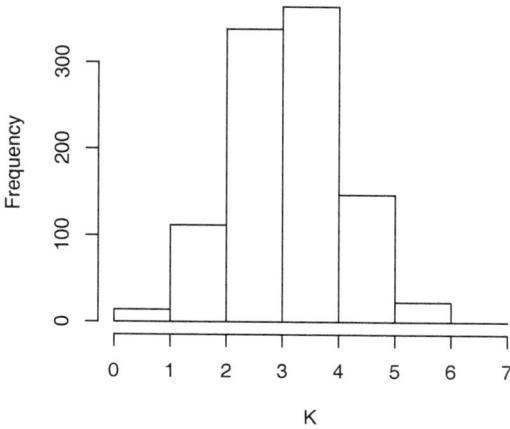

Fig. 6.16. Eye-Tracking example. Number of components as determined by KLMCMC.

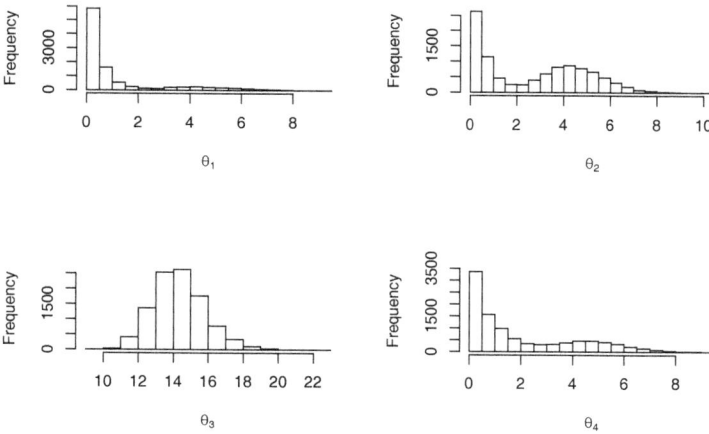

Fig. 6.17. Eye-Tracking example. Component means φ as determined by KLMCMC. Graph notation: In terms of Eq. (6.109), $\theta = \varphi$.

of the posterior distribution of parameters which appears in spite of the ordering of weights in stick-breaking settings. However, due to the stickiness of the Gibbs sampler, not all components are affected equally by label switching, see Figures 6.17 and 6.18 .

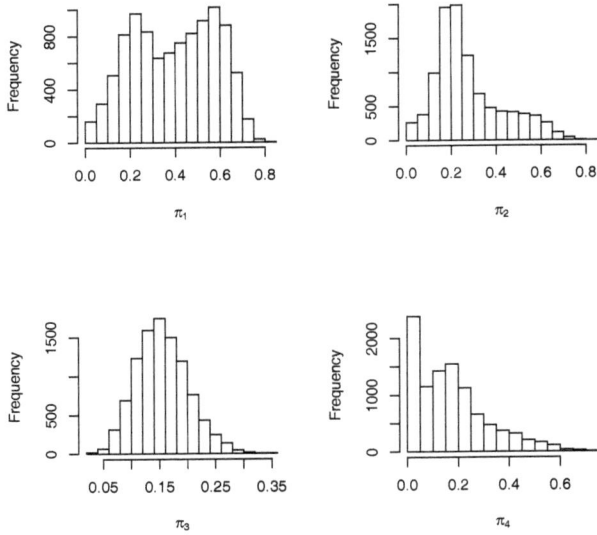

Fig. 6.18. Eye-Tracking example. Component weights w as determined by KLMCMC. Graph notation: In terms of Eq. (6.109), $\pi = w$.

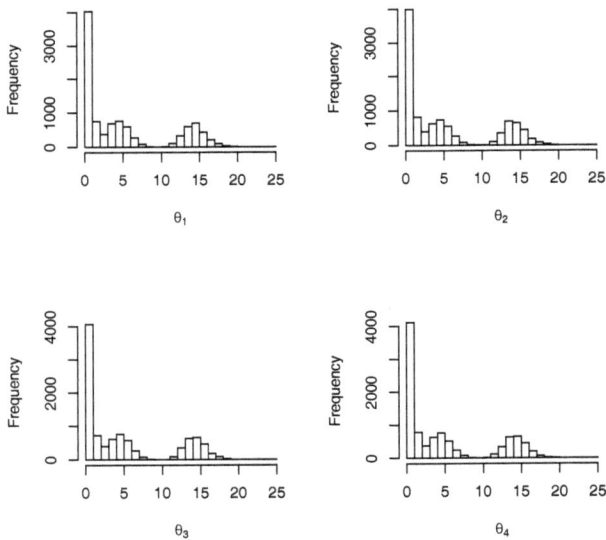

Fig. 6.19. Eye-Tracking example. Component means φ as determined by KLMCMC with random permutation sampling, conditional on $K = 4$. Graph notation: In terms of Eq. (6.109), $\theta = \varphi$.

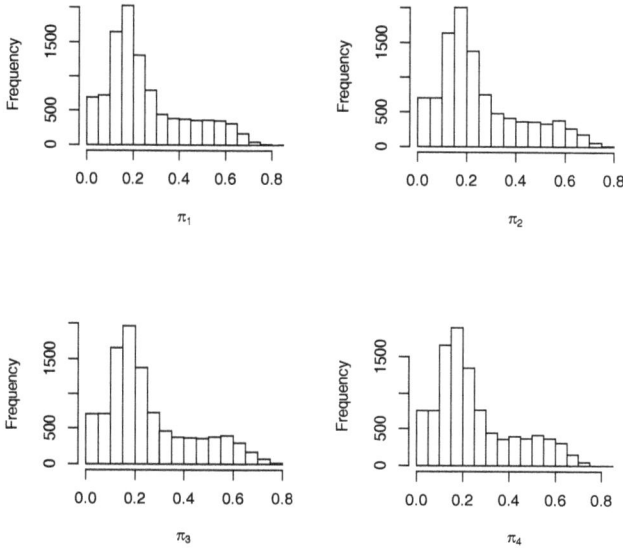

Fig. 6.20. Eye-Tracking example. Component weights w as determined by KLMCMC with random permutation sampling, conditional on $K = 4$. Graph notation: In terms of Eq. (6.109), $\pi = w$.

In order to improve posterior estimates of the model parameters, it is necessary to use KLMCMC with the random permutation sampler (Figures 6.19 and 6.20) followed by Stephens' relabeling.

6.8.5 *Stephens' relabeling for Poisson/gamma case*

Stephens' relabeling algorithm (Algorithm 3.3) from Section 3.5 can be applied in the Poisson/gamma case. For the purpose of relabeling, the model can be written as

$$p(Y_i|\Psi) = \sum_{k=1}^{K} w_k Pois(Y_i|\varphi_k),$$

$$\varphi_k \sim G(\cdot|1, 1), \tag{6.113}$$

where $\{w_k\}$ are the stick-breaking weights defined in Eq. (6.96), with $c = 1$, and where $\Psi = (w_1, ..., w_K; \varphi_1, .., \varphi_K)$.

Algorithm 6.1 Relabeling algorithm for Poisson/gamma case

Require: Start with some initial permutations $\nu_1, ..., \nu_T$ (e.g. the identity). Assume that $\Psi^{(t)}, t = 1, ..., T$ are sampled values of Ψ from a Markov chain with stationary distribution $p(\Psi|Y)$. Then iterate the following steps until a fixed point is reached.

Ensure: 1: Compute $\hat{q}_{ij}$:

$$\hat{q}_{jk} = \frac{1}{T} \sum_{t=1}^{T} p_{jk}(\nu(\Psi^{(t)})),$$

$$(6.114)$$

$$\text{where } p_{jk}(\Psi) = \frac{w_k Pois(Y_j|\varphi_k)}{\sum_{m=1}^{K} w_m Pois(Y_j|\varphi_m)}.$$

Ensure: 2: For $t = 1, ..., T$ choose ν_t to minimize

$$D_t = \sum_{j=1}^{N} \sum_{k=1}^{K} p_{jk}(\nu_t(\Phi^{(t)})) log \frac{p_{jk}(\nu_t(\Phi^{(t)}))}{\hat{q}_{jk}}.$$

$$(6.115)$$

This can be accomplished by examining all $k!$ permutations for each ν_t $t = 1, ..., T$.

Table 6.8 and Figures 6.21, and 6.22 show results of the application of Stephens' relabeling algorithms to the Eye-Tracking example.

6.8.6 *Pharmacokinetics example*

We return to the important PK population analysis problem. So far, we have considered this problem for all the methods previously discussed in this book. Now we treat this problem by the truncated stick-breaking method.

Consider administering a drug to a group of N patients and taking a series of T observations of the drug level in the blood. We have simulated measurements for 100 "patients" using the following PK model to describe

Table 6.8   Eye-Tracking example. Posterior estimations of model parameters, before and after relabeling.

	φ_1	φ_2	φ_3	φ_4	w_1	w_2	w_3	w_4
Before	5.04	5.08	5.06	5.00	0.25	0.25	0.25	0.25
After	0.23	1.00	4.60	14.34	0.39	0.26	0.20	0.15

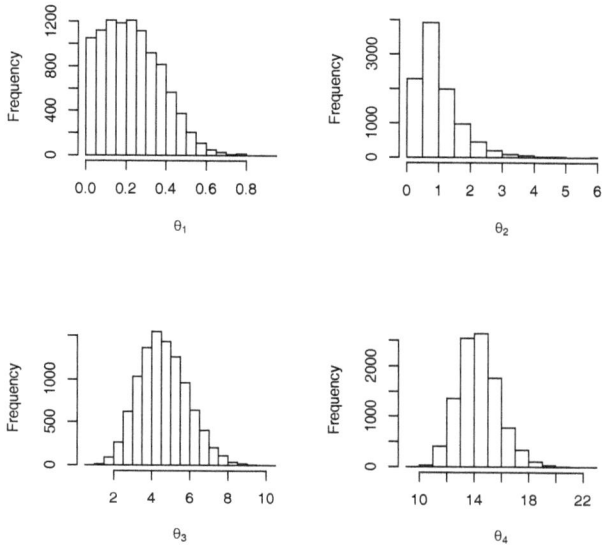

Fig. 6.21. Eye-Tracking example. Component means φ, conditional on $K = 4$, after Stephens' relabeling. Graph notation: In terms of Eq. (6.109), $\theta = \varphi$.

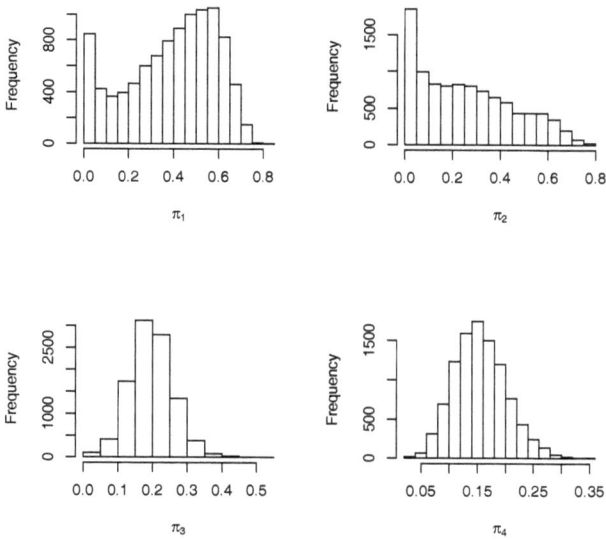

Fig. 6.22. Eye-Tracking example. Component weights w, conditional on $K = 4$, after Stephens' relabeling. Graph notation: In terms of Eq. (6.109), $\pi = w$.

the measured serum concentrations:

$$Y_{ij} = \frac{D}{V_i} e^{-\kappa_i t_j} + \sigma_e,$$

$$\kappa_i \sim \sum_{k=1}^{2} w_k N(\cdot | \mu_k, \sigma^2), \ V_i \sim N(\cdot | V_0, \sigma_v^2), \tag{6.116}$$

$$i = 1, \cdots, N, \ j = 1, \cdots, T.$$

In the framework of this model $\theta = (\kappa, V)$; the elimination rate constant κ is bimodal and the volume of distribution V is unimodal. True values of the model parameters are $V_0 = 20$, $w_1 = 0.8$, $w_2 = 0.2$, $\sigma_e = 0.1$, $D = 100$, $N = 100$, $T = 5$, $\mu_1 = 0.3$, $\mu_2 = 0.6$, $\sigma = 0.06$, $\sigma_v = 2$, $t_j \in \{1.5, 2, 3, 4, 5.5\}$.

In the truncated stick-breaking representation of Eq. (6.96), the model will be

$$Y_{ij} \sim N\left(\cdot | \frac{D}{V_i} e^{-\kappa_i t_j}, \sigma_e^2\right),$$

$$(\kappa_i, V_i) \sim \sum_{k=1}^{K} w_k \delta_{\varphi_k}(\cdot), \tag{6.117}$$

where $\varphi_k = (\tilde{\kappa}_k, \tilde{V}_k)$, and K is sufficiently large.

$$P(Y_{ij}|F) = \sum_{k=1}^{K} w_k N\left(\cdot | \frac{D}{\tilde{V}_k} e^{-\tilde{\kappa}_k t_j}, \sigma_e^2\right). \tag{6.118}$$

The base measure G_0 is given by

$$\tilde{\kappa}_k \sim N(\cdot | \kappa_0, 0.1), \ \kappa_0 \sim N(\cdot | 0.5, 0.1),$$

$$\tilde{V}_k \sim N(\cdot | V_0, 0.25), \ V_0 \sim N(\cdot | 20, 1), \tag{6.119}$$

and the precision parameter is

$$c \sim G(\cdot | 0.1, 0.1). \tag{6.120}$$

The model was implemented in JAGS [Plummer (2003)] using the RJAGS R package. We used one Markov chain for the Monte Carlo simulation, drawing every 10th posterior sample from iterations 10,000 to 20,000. The posterior distribution of the number of support points is shown in Figure 6.23. The posterior distributions of volume of distribution (Vol) and elimination constant (Kel) are shown as gray bars in Figure 6.24. To address the question of how many distinct groups there are in the studied

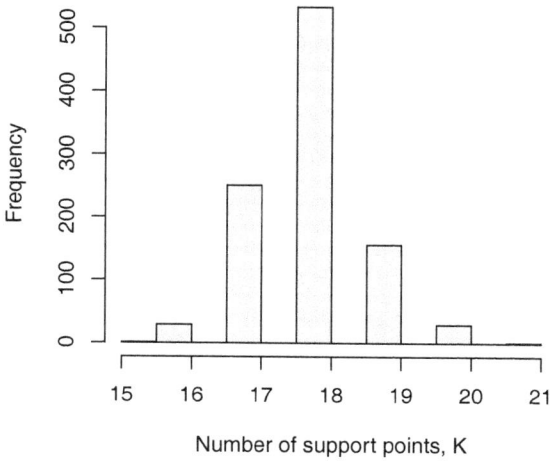

Fig. 6.23. Pharmacokinetics example: Posterior distribution of the number of support points.

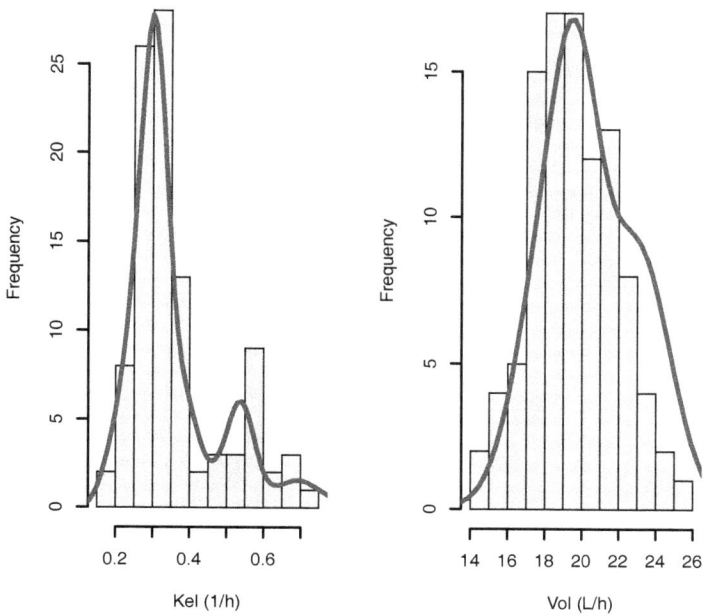

Fig. 6.24. Pharmacokinetics example. Marginal distribution of the elimination constant (left) and volume of distribution (right).

population, we need to generate "smoothed" distributions of Vol and Kel from the histograms. To achieve this, every support point is fitted with a normal distribution with mean at the support point and empirically selected standard deviation. In our case we selected the standard deviation for the smoothing distribution to be 0.05. The height of each normal peak is proportional to the "weight" of the support point. When all distributions are added up, we obtain a smooth curve (examples for Vol and Kel are shown in solid lines in Figure 6.24). Using these smoothed distributions, the number of components is easy to interpret. For example, we can be confident that there are two peaks in the distribution of Kel and one peak in the distribution of Vol.

6.9 Maximum Likelihood and Stick-Breaking (A Connection Between NPML and NPB Approaches)

This example is taken from the PhD dissertation of Alyona Chubatiuk [Chubatiuk (2013)]. It shows a connection between the methods of NPML and NPB.

Let $F_K^{(t)} = \sum_{k=1}^{K} w_k^{(t)} \delta(\phi_k^{(t)})$ be the distribution at the tth iteration of the blocked Gibbs sampler (GS) algorithm.

(1) The likelihood of $F_K^{(t)}$ is

$$L(F_K^{(t)}) = p(Y_1, ..., Y_N | F_K^{(t)}) = \prod_{i=1}^{N} p(Y_i | F_K^{(t)})$$

$$= \prod_{i=1}^{N} \int p(Y_i | \theta) dF_K^{(t)} = \prod_{i=1}^{N} \sum_{k=1}^{K} w_k^{(t)} p(Y_i | \phi_k^{(t)}), \quad (6.121)$$

where the last equality is true since $F_K^{(t)}$ is a finite sum of weighted delta functions. Then the log-likelihood can be written as follows:

$$l(F_K^{(t)}) = log(L(F_K^{(t)})) = log \left(\prod_{i=1}^{N} \sum_{k=1}^{K} w_k^{(t)} p(Y_i | \phi_k^{(t)}) \right)$$

$$= \sum_{i=1}^{N} log \left[\sum_{k=1}^{K} w_k^{(t)} p(Y_i | \phi_k^{(t)}) \right]. \quad (6.122)$$

(2) If we let $l_{max}(F_K^{(t)}) = \max_t(l(F_K^{(t)}))$ then we can compare $l_{max}(F_K^{(t)})$ to the value of $log(F^{ML})$. Since F^{ML} maximizes the likelihood function, the following inequality is always true:

$$l_{max}(F_K^{(t)}) \leq log(F^{ML}). \qquad (6.123)$$

Note that equality is possible with a big enough number of iterations of the blocked GS. We next show that this inequality is satisfied for the Galaxy dataset.

6.9.1 *Galaxy dataset*

Model:

$$Y_i = \phi_{S_i} + e_i, \text{ where } i = 1, ..., 82,$$

$$e_i \sim N(\cdot|0, \sigma^2), \ S_i \sim \sum_{k=1}^{K} w_k \delta_k,$$

$$\phi_k \sim G_0 = N(\cdot|\mu_0, \sigma_0^2), \qquad (6.124)$$

$$v_k \sim Beta(1, \alpha_0 = 1), \text{ where } k = 1, ..., K - 1; \ v_K = 1,$$

$$w_1 = v_1, w_k = (1 - v_1)(1 - v_2) \cdots (1 - v_{k-1})v_k, \ k = 2, ..., K.$$

We ran the blocked GS for 10,000 iterations for the following values of the parameters: $\sigma^2 = 1$, $\mu_0 = 20.8315$ sample mean and $\sigma_0^2 = (4.5681)^2$ sample variance. The maximum value of the log-likelihood that we obtained was $l_{max}(F_K^{(t)}) = -200.82$. Next, the value of the log-likelihood function of F^{ML} that we found by using the NPAG program of Section 6.2.4 was $l(F^{ML}) = -199.49$.

We can see that Eq. (6.122) is true in this example and, moreover, the values are very close. In the two figures below we present the graphs of F^{ML} and $F_K^{(opt)} = argmax(F_K^{(t)})$.

The distributions for F^{ML} and $F_K^{(opt)}$ are very similar (see Figure 6.25), even though they are calculated by different means. However, the likelihoods are very close, which then gives another method for approximating the maximum likelihood distribution, and it works both ways.

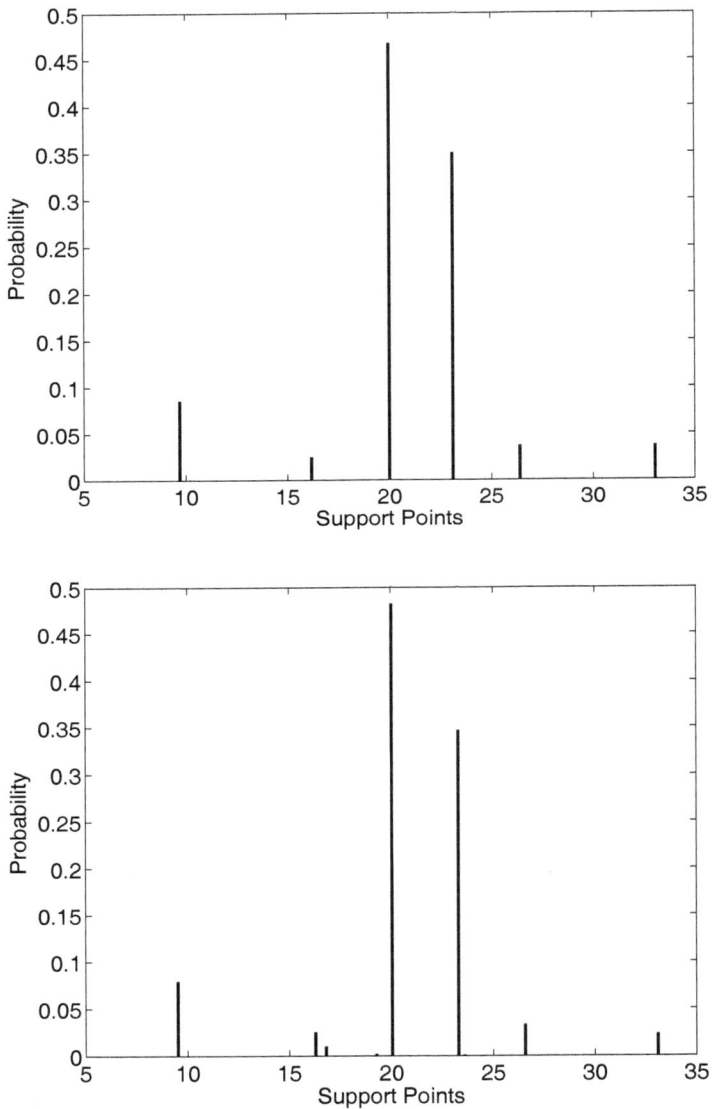

Fig. 6.25.   Galaxy dataset. Top: the graph of F^{ML}. There are six support points for the NPML method. Bottom: the graph of $F_K^{(opt)}$. There are seven support point for the NPB method.

Chapter 7

Bayesian Clustering Methods

In this chapter, we discuss the application of algorithms developed in the previous chapters to different datasets and present our Kullback–Leibler clustering (KLC) method for the treatment of data-rich problems. We start with a description of previously developed clustering methods and the motivation for the development of this approach (Section 7.1). We show the application of KLMCMC (KL Markov chain Monte Carlo) to gene expression data (Section 7.2). Next, we describe a novel algorithm for selection of the optimal number of components in Section 7.3, and demonstrate utility of this method (Section 7.4). We conclude this chapter with the application of the nonparametric Bayesian (NPB) method to the prediction of transcription start sites (Section 7.5).

7.1 Brief Review of Clustering Methods in Microarray Analysis

Gene clustering analysis is important for finding groups of correlated (potentially co-regulated) genes. A number of clustering methods have been developed specifically with gene expression in mind, in addition to numerous methods adapted from other disciplines. These methods include the following popular approaches: hierarchical clustering [Eisen *et al.* (1998)], self-organizing maps [Kohonen (1990); Tamayo *et al.* (1999)], K-means [MacQueen (1967); Hartigan and Wong (1979)], principal component analysis (PCA) [Raychaudhuri *et al.* (2000); Yeung and Ruzzo (2000)], singular value decomposition (SVD) [Alter *et al.* (2000)], partitioning around medioids (PAM) [Kaufman and Rousseeuw (1990)], model-based clustering [Yeung *et al.* (2001); Fraley and Raftery (2007, 2002); McLachlan *et al.* (2002); Medvedovic and Sivaganesan (2002); Medvedovic *et al.* (2004); Barash and Friedman (2002)], tight clustering [Tseng and Wong (2005)], and curve clustering [Wakefield *et al.* (2003); De la Cruz-Mesía *et al.* (2008)].

As was recently demonstrated in Thalamuthu *et al.* (2006), model-based and tight clustering algorithms consistently outperform other clustering methods. Tight clustering and model-based clustering make provision for the existence of a set of genes, unaffected by the biological process under investigation. This set of genes is not clustered, and thus false positive cluster members that almost inevitably appeared in traditional methods do not distort the structure of identified clusters. Many traditional clustering methods based on gene expression produce groups of genes that are co-expressed but not necessarily co-regulated [Clements *et al.* (2007)]. The rapidly growing number of fully sequenced and annotated genomes and diverse datasets describing the transcriptional process calls for development of combined probability models based on regulatory elements and gene expression. Integrative approaches, developed by Clements *et al.* (2007), Barash and Friedman (2002), Jeffrey *et al.* (2007) and others, provide deeper insight into cellular processes.

Another improvement of the model-based clustering methods is utilization of a biological model. Traditional measures of similarity between genes, e.g. correlation coefficients, Euclidian distance, Kendal's tau and city-block distance, report a single number for a pair of genes, not utilizing the data-rich nature of time-series microarray experiments. Model-based clustering methods [e.g. Wakefield *et al.* (2003); Yeung *et al.* (2001); Fraley and Raftery (2002); Fraley and Raftery (2007); Medvedovic and Sivaganesan (2002); Medvedovic *et al.* (2004); McLachlan *et al.* (2002); De la Cruz-Mesía *et al.* (2008)] are able to take advantage of this situation. We also note that Medvedovic and Sivaganesan (2002) used a Dirichlet process mixture model for clustering.

The widely used program MCLUST, developed by Fraley and Raftery, provides an iterative expectation-maximization (EM) method for maximum likelihood clustering for parameterized Gaussian mixture models. According to Fraley and Raftery, each gene is represented by a vector in a space of arbitrary dimension d, and interaction between genes is given in the form of a d-dimensional covariance matrix. In 2003, Wakefield *et al.* (2003) described a curve-clustering approach and used a Bayesian strategy to partition genes into clusters and find parameters of mean cluster curves. One crucial difference between the approach of Wakefield and Zhao and previously developed methods is that neither Yeung *et al.* (2001) nor Fraley and Raftery (2002, 2007) acknowledged the time-ordering of the data. Recently Tatarinova (2006), Tatarinova *et al.* (2008) and independently Li (2006) and De la Cruz-Mesía *et al.* (2008) developed model-based clustering methods that group individuals according to their temporal profiles of gene expression using nonlinear Bayesian mixture models and maximum likelihood approaches. These approaches reconstruct underlying time dependence for each gene and provide models for measurement errors and noise. The goal is to cluster the parameters of nonlinear curves, not the observed values. Hence these methods can efficiently handle missing data as well as irregularly spaced time points. Model-based clustering has some inherent advantages compared to non-probabilistic clustering techniques [Li (2006)]. Probabilistic model-based clustering methods provide estimations of the distribution of parameters for each cluster as well as posterior probabilities of cluster membership.

Fraley and Raftery (2002) demonstrated the effectiveness of model-based clustering for medical and gene expression data. As was pointed out by Thalamuthu *et al.* (2006), model-based clustering methods enjoy full probabilistic modeling and the selection of the number of clusters is statistically justified. Due to the complexity of large-scale probabilistic modeling, they suggest that model-based clustering can be successfully performed as higher-order machinery that can be built upon traditional clustering methods. Pritchard *et al.* (2000) developed clustering methods to infer population structure and assign individuals to populations using multi-locus genotype data. The method was based on Stephens' (2000a) approach using Markov chain Monte Carlo techniques.

In the next section we discuss the KLMCMC method as one of many possible higher-order clustering methods.

7.2 Application of KLMCMC to Gene Expression Time-Series Analysis

The temporal program of sporulation in budding yeast was studied in Chu *et al.* (1998). The dataset consists of seven successive time points ($t = 0, 0.5, 2, 5, 7, 9, 11.5$ hours). We used the pre-screening strategy of Wakefield *et al.* (2003) and used the first time point $t = 0$ to estimate measurement errors. According to the analysis of Chu *et al.* (1998), and to the false discovery rate filter by Zhou (2004) and Wakefield *et al.* (2003), approximately 1,300 of 6,118 genes showed significant changes in mRNA level in

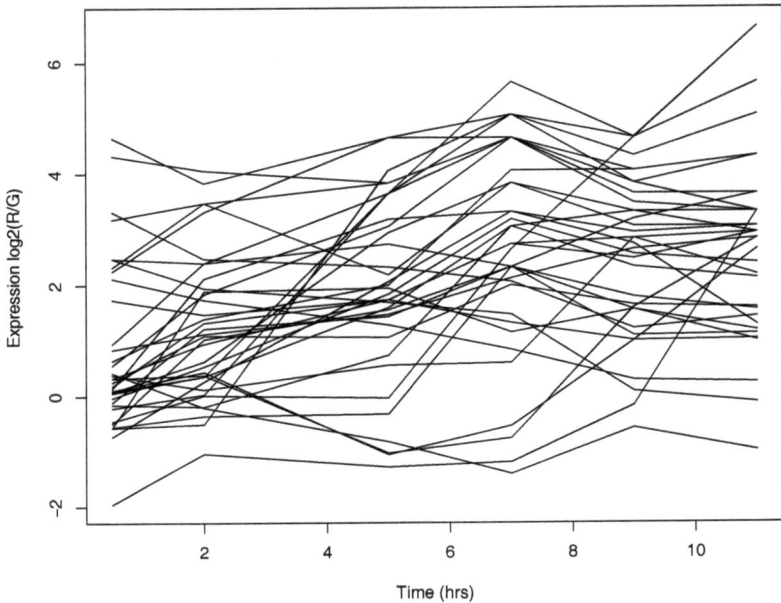

Fig. 7.1.   Gene expression time-series of sporulation in budding yeast, corresponding to seven characteristic profiles, as identified in Chu *et al.* (1998).

the course of the experiment. Wakefield *et al.* (2003) assumed a first-order random walk model for gene-specific trajectories. Chu *et al.* (1998) had identified seven characteristic profiles and grouped genes by similarity to these profiles:

(1) Metabolic: induce rapidly and transiently after the transfer to sporulation medium.
(2) Early (I): detectable 0.5 hours after transfer to sporulation medium and sustained expression through the rest of the time course. Role in recombination and chromosome pairing.
(3) Early (II): delayed increase in transcription level.
(4) Early-middle: initially induced 2 hours after the transfer to sporulation medium and in addition at 5 and 7 hours.
(5) Middle: expressed between 2 and 5 hours.
(6) Middle-late: from 5 to 7 hours, involved in meiotic division.
(7) Late: induced between 7 and 11.5 hours, spore wall maturation.

For our analysis we assume that the temporal patterns of the studied genes can be described by a function $f(\theta_i, t_j) = P_i(t_j)e^{\delta_i t_j}\eta(t_j - \gamma_i)$, where $P_i(t_j) = \alpha_i + \beta_i t_j$ is a first-degree polynomial and $\theta = \{\alpha_i, \beta_i, \delta_i, \gamma_i\}$ is a set of gene-specific parameters. In order to represent the time γ_i when the ith gene is "turned on" we use the step function $\eta(t - \gamma)$:

$$\eta(t - \gamma) = \begin{cases} 0, & \text{if } t < \gamma; \\ 1, & \text{if } t \geq \gamma. \end{cases} \tag{7.1}$$

Observations are modeled as $y_{ij} \sim N(\cdot | f(\theta_i, t_j), \sigma_e^2)$ for $i = 1, ..., N$ and $j = 1, ..., T$ where N is the number of genes, $T = 7$ is the number of experiments, and $\sigma_e^{-2} \sim G(\cdot | g, h)$ is the experimental error. Values of g and h are estimated from the $T = 0$ observations. Gene-specific parameters $\{\alpha_i, \beta_i, \delta_i, \gamma_i\}$ are assumed to have a multivariate normal mixture distribution:

$$(\alpha_i, \beta_i, \delta_i, \gamma_i) \sim \sum_{k=1}^{N} w_k N(\cdot | \mu_k, \Sigma_k). \tag{7.2}$$

Component weights are assumed to have the Dirichlet distribution $w \sim Dir(\cdot | \alpha)$. Least-squares estimates are used to obtain the parameters μ_0, Σ_0 of weakly informative priors:

$$\mu_k \sim N(\cdot | \mu_0, \Sigma_0) \tag{7.3}$$

for parameters α_i, β_i and δ_i; the prior distribution for the "turn-on time" γ_i is assumed to have a uniform distribution $\gamma_i \sim U[0, 12]$. To simplify the computation, we have assumed that the covariance matrices Σ_k and Σ_0 are diagonal: $[\Sigma_k^{-1}]_{ll} \sim G(a_l, b_l)$, for $l \in [1, 3]$, where the parameters a_l, b_l of the gamma distributions are the least-squares estimates. For the benchmark we used a subset of genes involved in the process of sporulation, containing $N = 35$ well-characterized genes.

Using the KLMCMC method, the dimension-changing step consists of three parts, as defined in Section 4.4.

(1) The death rate is estimated as $\delta(Q) = \sum_{k=1} 1/d^*(k)$, where $d^*(k)$ is the smallest weighted KL distance between component k and all other components. It is computed analytically in the case of normal distributions and evaluated by the Gibbs sampler based on model parameters. Birth or death move is chosen with probabilities $p_b = \frac{\lambda_B}{\lambda_B + \delta(Q)}$ and $p_d = \frac{\delta(Q)}{\lambda_B + \delta(Q)}$.

(2) Labels of components are randomly permuted.

Table 7.1   Yeast sporulation time series. KLMCMC profiles conditional on $K = 7$.

Cluster	Equation, $f_{cluster}(t)$
Metabolic	$(1.1954 + 0.6235t)e^{-0.2105(t-0.2585)}\eta(t - 0.2585)$
Early I	$(3.034 + 0.5189t)e^{-0.06239(t-0.2526)}\eta(t - 0.2526)$
Early II	$(0.1588 + 0.7532t)e^{-0.1678(t-1.174)}\eta(t - 1.174)$
Early-middle	$(0.2866 + 0.6336t)e^{-0.04974(t-0.7561)}\eta(t - 0.7561)$
Middle	$(-0.01791 + 0.7208t)e^{-0.09469(t-3.489)}\eta(t - 3.489)$
Middle-late	$(-0.0074 + 0.3987t)e^{0.00197(t-5.958)}\eta(t - 5.958)$
Late	$(-0.4577 + 0.2184t)e^{0.08551(t-7.632)}\eta(t - 7.632)$

(3) Initial values of parameters for the next round of Gibbs sampler are calculated from the permuted output of the previous run.

As was previously reported in Zhou and Wakefield (2005) and Zhou (2004), the number of clusters is highly sensitive to the choice of prior distributions. Hence, biological and experiment-specific insight should be used to determine informative priors.

Based on the biological analysis in Chu *et al.* (1998), we set the expected number of clusters $\lambda_B = 7$ and ran the simulation for 1,000 hybrid birth–death steps, each containing 5,000 WinBUGS iterations, where the first 4,000 WinBUGS iterations were discarded as "burn-in". Stephens' relabeling method (Algorithm 3.3) was used to determine parameters of individual clusters and cluster membership.

The number of mixture components varied from $K = 4$ to $K = 10$. The Markov chain spent most (34%) of the time in the state $K = 7$, 26% of the time in the state $K = 6$, and 18% in the state $K = 8$. We used the parameters estimated for $K = 7$ to describe the KLMCMC profiles for individual clusters (Table 7.1). It is remarkable that the KLMCMC approach was able to reproduce the results of Chu *et al.*, who partitioned the genes involved in the yeast sporulation process into exactly seven groups. KLMCMC efficiently dealt with observation errors and the gene-specific variances.

KLMCMC profiles are shown in Figure 7.2. KLMCMC produced a smooth and biologically meaningful description of gene expression time dependence.

Clusters obtained by the KLMCMC process can be further extended using the entire collection of microarray data. For example, late sporulation genes are described by the KLMCMC profile as

$$f_{late}(t) = (-0.4577 + 0.2184t)e^{0.08551(t-7.632)}\eta(t - 7.632). \tag{7.4}$$

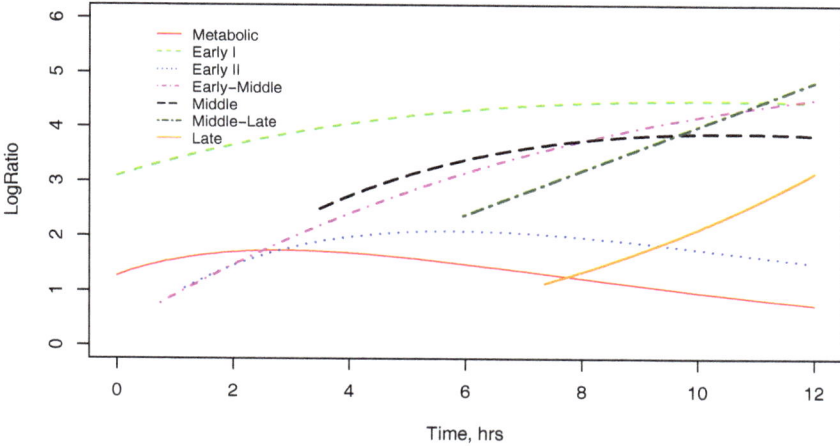

Fig. 7.2.   Seven clusters of temporal patterns of sporulation in budding yeast, as defined by KLMCMC, see Table 7.1.

All un-clustered genes are ordered by Pearson's correlation coefficient of expression profiles and compared to the cluster in a sequential manner. After an addition of each gene, parameters of the cluster are re-computed. If a parameter distribution becomes too vague, upon inclusion of an extra gene, this gene is not added to a cluster. When we added 15 cluster members, the distribution of cluster parameters became

$$f_{late}^{(15)}(t) = (-0.322 + 0.141t)e^{0.09(t-7.72)}\eta(t - 7.72). \qquad (7.5)$$

In this manner we can add as many as 71 genes without distorting the initial parameter distributions:

$$f_{late}^{(71)}(t) = (-0.21 + 0.12t)e^{0.04(t-6.1)}\eta(t - 6.1), \qquad (7.6)$$

shown in Figure 7.3. If we compare this approach with the traditional nearest neighbor clustering based on Pearson's' correlation coefficient of expression profiles, the cluster mean for the top 71 genes with correlation coefficient above 0.9 will be shifted towards negative expression values (Figure 7.3).

7.3   Kullback–Leibler Clustering

In this section we suggest a new method for the clustering of data-rich time-series observations: *Kullback–Leibler Clustering* (KLC). Our approach

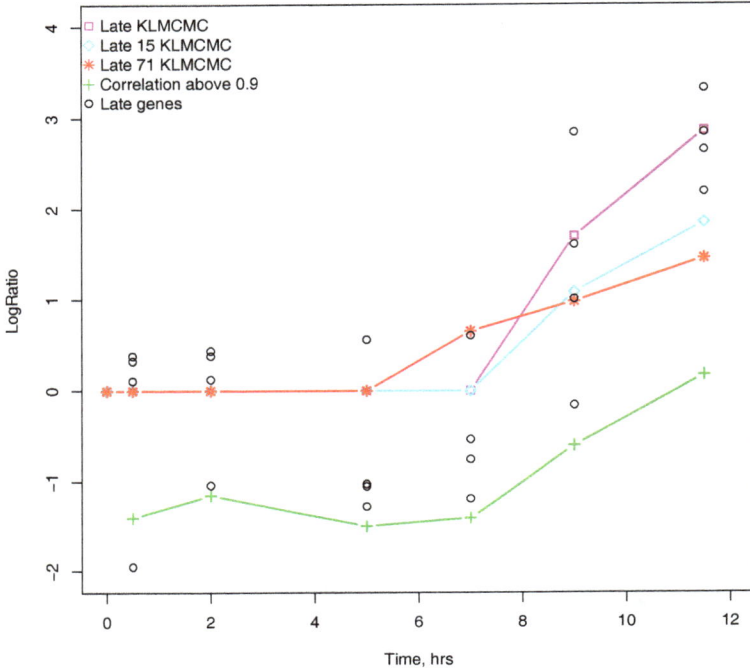

Fig. 7.3. Late KLMCMC cluster, extended "late" clusters with 15 and 71 members, and the nearest neighbor clustering results for the top 71 genes, with expression highly correlated to the members of the "late" sporulation cluster.

assumes that every cluster can be described by a smooth centroid (mean) curve. If N genes can be grouped into K groups by their expression profile, then all individuals (i.e. genes or patients) that belong to a group $k = 1, ..., K$ have similar values of the gene-specific "trajectory" parameters θ, where θ is an n-dimensional vector. For each cluster k, the parameter vector $\theta^{(i)}$, $i \in 1, ..., n$ is assumed to be normally distributed $N(\mu_k, \Sigma_k)$, where, in the simplest case, the variance-covariance matrices Σ_k are assumed to be diagonal.

The KLC algorithm is implemented as agglomerative clustering. The approach is based on the extension of the idea of Sahu and Cheng (2003) to use the weighted KL distance to find the optimal number of mixture components and is feasible in data-rich situations. The difference between the method of Sahu and Cheng (2003) and our method is that we suggest to iteratively collapse multiple components following only one run of the

Gibbs sampler. The KLC shortcut provides a significant computational advantage and speedup over KLMCMC for the clustering of large datasets.

At the first step of the KLC method, we assume that all genes are allocated to separate clusters (components). For every individual $i = 1, ..., N$, values of θ_i can be found by analyzing a model in WinBUGS. The KLC method is applicable for the solution of several possible nonlinear models; one of the many possibilities is

$$p(y_{ij}|\theta_i, t_j, \tau) = N(y_{ij}|f(t_j, \theta_i), \tau^{-1}),\qquad(7.7)$$

$$\theta_i \sim N(\cdot|\mu_i, \Sigma_i),$$

$$\tau \sim G(\cdot|a, b),$$

$$\mu_i \sim N(\cdot|\eta, C),$$

$$\Sigma_i^{-1} \sim W(\cdot|q, \Psi),$$

where $j = 1, ..., T$ and $f(t_j, \theta_i)$ is some nonlinear function. For computational simplicity we assume independence of the parameter vector components: for all $i = 1, ..., N$, the covariance matrix Σ_i is diagonal. This assumption greatly increases the speed of computation, but it is not necessary from the theoretical point of view. Note, that Eq. (7.7) is not a mixture model. The fitting of these equations requires a data-rich situation $T \gg dim(\theta_i)$. At convergence, the posterior means of $\mu_1, ..., \mu_N$ and $\Sigma_1, ..., \Sigma_N$ are obtained, and denoted by the same symbols.

At the second step we construct a "pseudo-mixture model" using parameter values estimated in Step 1.

$$f^{(N)}(\theta) = \sum_{i=1}^{N} w_i N(\theta|\mu_i, \Sigma_i), w_i = \tfrac{1}{N} \text{ for all } i,\qquad(7.8)$$

where each component of a mixture corresponds to an individual. As a measure of dissimilarity between individuals/genes h and g we will use distance $\tilde{d}(h, g)$ defined in Eq.(4.41). For the cut-off condition we will use

$$\tilde{D} = \sum_{i=1}^{n} \sum_{m=1}^{M} d_m^{(i)} = \sum_{i=1}^{n} \sum_{m=1}^{M} \frac{k_m}{2K} log \left(1 + \frac{|\delta\mu^{(i)}|^2}{2\sigma^{(i)2}}\right) \leq \frac{nlog(3)}{2}\qquad(7.9)$$

It is necessary to compute $\binom{N}{2}$ distances,

$$\tilde{d}(h,g) = \frac{1}{2}\{d(f(\mu_h, \Sigma_h), f(\mu^*_{h,g}, \Sigma^*_{h,g})) + d(f(\mu_g, \Sigma_g), f(\mu^*_{h,g}, \Sigma^*_{h,g}))\},$$
(7.10)

to be used as input for a hierarchical clustering procedure. Hierarchical clustering is a method of data partitioning resulting in a hierarchy of clusters. The results of hierarchical clustering are usually presented in a *dendrogram*, also known as *cluster tree*.

Fuest *et al.* (2010) provided the following concise description of this method. "Dendrograms graphically present the information concerning which observations are grouped together at various levels of dissimilarity. At the bottom of the dendrogram, each observation is considered its own cluster. Vertical lines extend up for each observation, and at various dissimilarity values, these lines are connected to the lines from other observations with a horizontal line. The observations continue to combine until, at the top of the dendrogram, all observations are grouped together. The height of the vertical lines and the range of the dissimilarity axis give visual clues about the strength of the clustering. Long vertical lines indicate more distinct separation between the groups. Long vertical lines at the top of the dendrogram indicate that the groups represented by those lines are well separated from one another. Shorter lines indicate groups that are not as distinct."

The specific *linkage criterion* used in clustering determines the distance between sets of observations as a function of the pairwise distances between observations. The two most common linkage criteria are *complete linkage* and *single linkage*. In the *single linkage* setting, the distance $D(A, B)$ between clusters A and B is derived from the pairwise distance between individuals as $D(A, B) = min_{a,b}d(a, b)$, where $a \in A, b \in B$. In the *complete linkage* setting, the distance is $D(A, B) = max_{a,b}d(a, b)$. At each iteration, the closest clusters A and B are combined.

The relationship between objects and clusters can be represented by a dendrogram (for example, see Figure 7.5), in which dissimilarity between two objects can be inferred from the height at which they are joined into a single cluster. The number of clusters can be determined using the similarity cut-off condition in Eq. (7.9).

In Algorithm 7.1 we use a complete linkage criterion to determine the distance between sets of observations as a function of the pairwise distances between observations.

Algorithm 7.1 Kullback–Leibler clustering algorithm

Require: Set the initial number of components $K = N$.

Ensure: 1: Assign all measurements to separate clusters (components) and fit the model. Run the Gibbs sampler (in WinBUGS, JAGS or Openbugs) for M iterations and determine the posterior estimates of the parameters (μ_i, Σ_i) for $i \in [1, ..., N]$.

Ensure: 2: Based on the Gibbs sampler output, calculate $\binom{N}{2}$ pairwise distances $\tilde{d}(h, g)$ according to Eq. (7.10) between all pairs of components.

Ensure: 3: Successively combine pairs of clusters.

repeat

For each component m find the nearest component m', so that $D(J_m, J_{m'}) = max_{a,b}\tilde{d}(a, b)$, where $a \in J_m, b \in J_{m'}$.

Compute $\tilde{D}$ according to Eq. (7.10).

until All components are merged into a single cluster.

7.3.1 *Pharmacokinetic example*

Consider the following PK model:

$$Y_{ij} = \frac{D}{V}exp(-\kappa_i t_j) + \sigma_e e_{ij}, \tag{7.11}$$

where $i = 1, ..., N$, $j = 1, ..., T$, and elimination constants κ_i are described as a mixture with an unknown number of K components:

$$\kappa_i = \sum_{k=1}^{K} w_k N(\cdot|\mu_k, \sigma_k^2). \tag{7.12}$$

Patients can belong to one of two groups according to their drug acetylation phenotype: fast acetylators (short drug half-life) or slow acetylators (long drug half-life). Slow acetylators are less efficient than fast acetylators in the metabolism of numerous drugs. For certain drugs, the acetylation polymorphism (both fast and slow) can be associated with an increased risk of toxicity [Blum *et al.* (1991)].

Using the model in Eq. (5.17), and assuming $D/V = 1$, $K = 2$, $\sigma_e = 0.1$, $\sigma_k = 0.06$, $w_1 = 0.8$, $w_2 = 0.2$, $\mu_1 = 0.3$ and $\mu_2 = 0.6$, "observations" were generated for $N = 100$ patients (Figure 7.4) and $T = 5$ time points $t_j = \{1.5, 2, 3, 4, 5.5\}$.

In the simplest case of one unknown parameter per subject, the number of observations is larger than the number of unknown parameters, and we

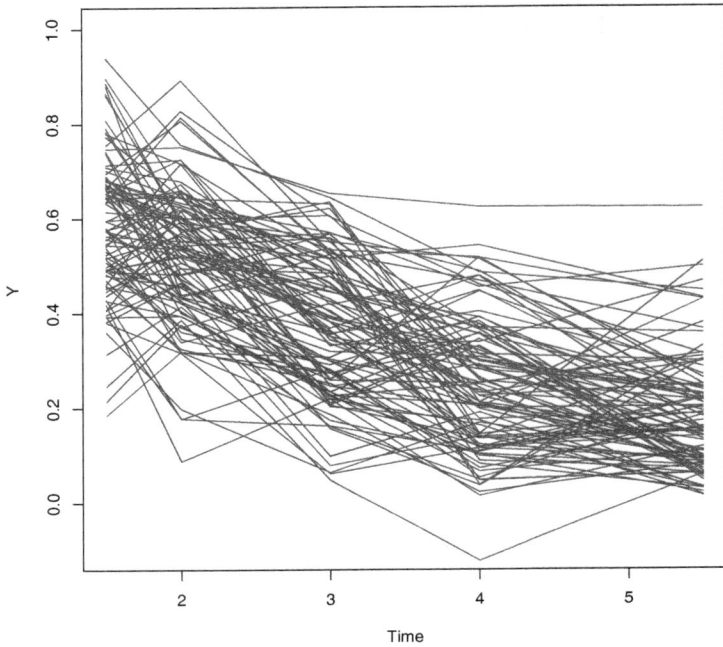

Fig. 7.4.   Pharmacokinetic simulated data.

used WinBUGS to analyze the following model:

$$Y_{ij} = exp(-\kappa_i t_j) + \sigma_e e_{ij},$$
$$\kappa_i \sim N(\mu_i, \sigma^2), \qquad\qquad (7.13)$$
$$\mu_i \sim N(\lambda_0, \sigma_0),$$

where known model parameters are $\sigma = 0.06$, $\lambda_0 = 0.36$, $\sigma_e = 0.1$. The Gibbs sampler was run for 20,000 iterations, with the first 10,000 iterations discarded as burn-in. The distance $\tilde{d}(h, g)$ between components corresponding to "patients" h and g, in this case, is reduced to:

$$\tilde{d}(h, g) = \frac{1}{N} log(1 + \frac{(\mu_g - \mu_h)^2}{4\sigma^2}). \qquad\qquad (7.14)$$

$\binom{100}{2}$ distances were used as input for hierarchical cluster analysis in R (function *hclust* from the package *stats*). The resulting KLC dendrogram is presented in Figure 7.5. For comparison, hierarchical clustering was repeated using $1 - r$ as an input distance, where r is Pearson's correlation

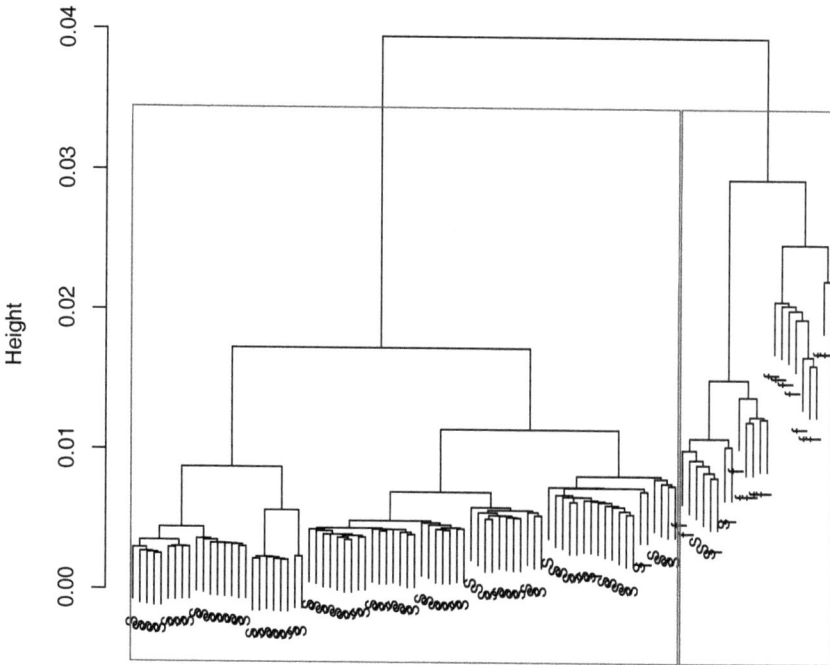

Fig. 7.5. Pharmacokinetic simulated data, cluster dendrogram for KLC.

coefficient. Pearson's correlation coefficient is generally used in the sciences as a measure of the strength of linear dependence between two variables. For two arrays of data $\{X\}$ and $\{Y\}$, Pearson's correlation coefficient is defined as $r = \frac{\sum_{i=1}^{n}(X_i-\bar{X})(Y_i-\bar{Y})}{\sqrt{\sum_{i=1}^{n}(X_i-\bar{X})^2}\sqrt{\sum_{i=1}^{n}(Y_i-\bar{Y})^2}}$. The corresponding dendrogram is shown in Figure 7.6. For both methods, borders were drawn around the two clusters. Based on the clustering results, all "patients" were divided into four groups:

True positive (TP) Slow acetylators correctly diagnosed as slow ones
False positive (FP) Fast acetylators incorrectly identified as slow ones
True negative (TN) Fast acetylators correctly identified as fast ones
False negative (FN) Slow acetylators incorrectly identified as fast ones

These counts are used to compute statistical measures of performance. Among the commonly used measures are:
$Sensitivity = \frac{TP}{TP+FN}$ (ability to identify positive results)

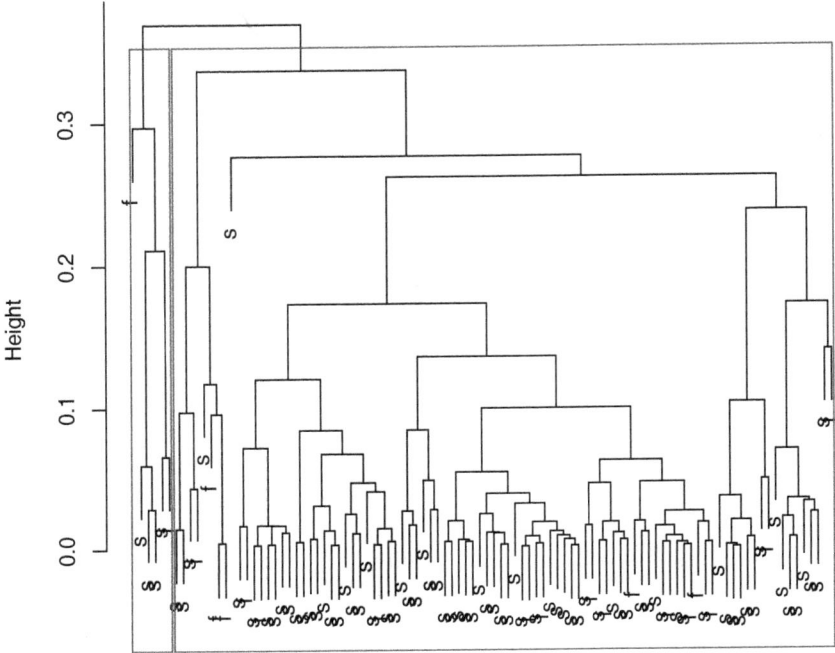

Fig. 7.6.   Pharmacokinetic simulated data, cluster dendrogram for hierarchical cluster-ing based on Pearson's correlation coefficient.

and
$Specificity = \frac{TN}{TN+FP}$ (ability to identify negative results).

In this example, KLC performed better than Pearson's correlation co-efficient for this noisy nonlinear dataset. For KLC:
$Sensitivity = \frac{TP}{TP+FN} = \frac{78}{78+2} = 0.975,$
$Specificity = \frac{TN}{TN+FP} = \frac{20}{20+0} = 1.$
While for Pearson's correlation coefficient:
$Sensitivity = \frac{TP}{TP+FN} = \frac{76}{76+4} = 0.95,$
$Specificity = \frac{TN}{TN+FP} = \frac{2}{2+18} = 0.1.$

Generally, if one can predict the type of the temporal pattern (e.g. exponential or polynomial) and the problem is data-rich, then model-based approaches, such as KLC have an advantage over correlation-coefficient based methods, which are sensitive to noise and the presence of outliers.

7.4 Simulated Time-Series Data with an Unknown Number of Components (Zhou Model)

In this section, we compare performance of the KLC algorithm with reductive stepwise method and KLMCMC using the Zhou example (page 89 in Zhou (2004)). We applied the three methods to the simulated dataset described by Zhou (2004) and compared the performance of these methods of choosing the optimal number of components. Following Zhou's approach, we generated 50 linear time curves from $K = 3$ clusters (30% of data points were generated from each of the first and second clusters, and 40% from the third cluster). The simulated data and time-series are shown in Figure 7.7 and Figure 7.8. Parameters of the clusters, namely, cluster centers and covariance matrices, for the three clusters were

$$\begin{cases} \mu_1 = \begin{pmatrix} 8 \\ 1 \end{pmatrix}, \Sigma_1 = \begin{pmatrix} 0.1 & 0 \\ 0 & 0.1 \end{pmatrix}, \\ \mu_2 = \begin{pmatrix} 10 \\ 0 \end{pmatrix}, \Sigma_2 = \begin{pmatrix} 0.1 & 0 \\ 0 & 0.1 \end{pmatrix}, \\ \mu_3 = \begin{pmatrix} 12 \\ -1 \end{pmatrix}, \Sigma_3 = \begin{pmatrix} 0.2 & 0 \\ 0 & 0.2 \end{pmatrix}. \end{cases} \quad (7.15)$$

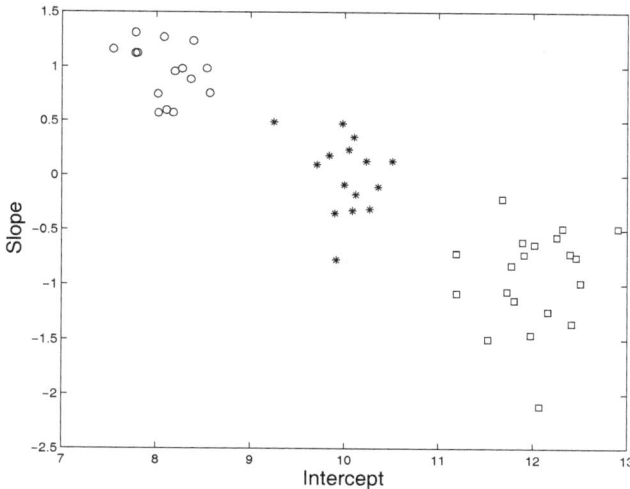

Fig. 7.7. Zhou model. Intercepts and slopes for three groups of simulated parameters for $K = 50$: group 1 – circle, group 2 – stars and group 3 – squares.

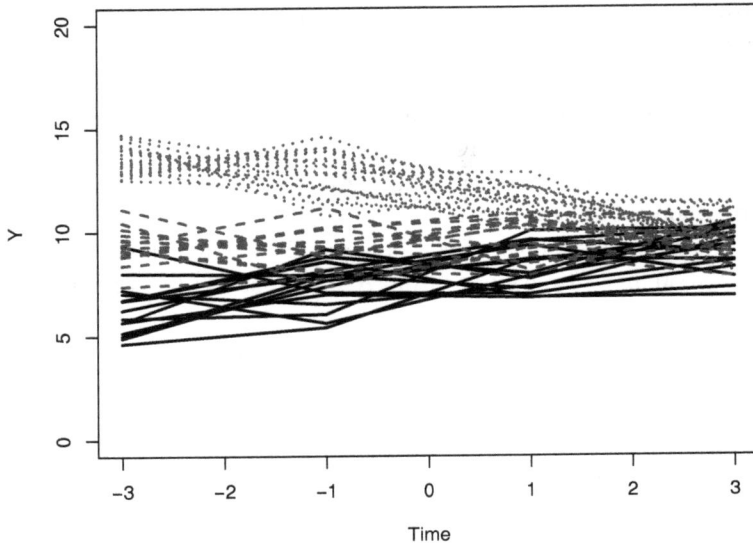

Fig. 7.8.   Zhou model. Simulated time series.

For each generated curve $i = 1, ..., 50$, intercept and slope were assumed to arise from the multivariate normal distribution:

$$\begin{pmatrix} \alpha_i \\ \beta_i \end{pmatrix} \sim N(\cdot | \mu_{Z_i}, \Sigma_{Z_i}), \tag{7.16}$$

where $Z_i \in \{1, 2, 3\}$ is a cluster membership label. There are four time points $t_j = (-3, -1, 1, 3)$ and observed data y_{ij} depend on the experimental error σ_e:

$$y_{ij} \sim N(\cdot | \alpha_i + \beta_i x_j, \sigma_e^2), \tag{7.17}$$

where α_i is an intercept and β_i is a slope for a curve i. Following Zhou (2004), we choose $\sigma_e = 1$. We will refer to this model as the *Zhou model*.

7.4.1   *Model description*

(1) The first stage model has the form: $y_{ij} \sim N(\cdot | \alpha_i + \beta_i t_j, \sigma_e^2)$, for $i = 1, ..., n, j = 1, ..., T$.

(2) The second stage describes intercept and slope of individual curves $(\alpha_i, \beta_i) = \theta_i$, where the θ_i arise from a mixture of K sub-populations, and where Σ_k is diagonal.

$$\theta_i | Z_i = k \sim N(\cdot | \mu_k, \Sigma_k), \text{ where } k = 1, ..., K. \tag{7.18}$$

(3) At the third stage we assume that cluster membership labels $\{Z_i\}$ are independent and follow the categorical distribution $p(Z_i = k | w_1 ... w_K) = w_k$.

(4) At the fourth stage we specify parameters:

$$(w_1, ..., w_K) \sim D(\cdot | \alpha_1, ..., \alpha_K),$$

$$\mu_k \sim N(\cdot | \eta, C), \tag{7.19}$$

$$\Sigma_k^{-1} \sim Wishart(\cdot | \rho, (\rho R)^{-1}).$$

(5) The fifth stage of the model specifies distribution of the experimental error:

$$\sigma_e^{-2} \sim G(\cdot | \lambda, \nu). \tag{7.20}$$

Hyper-parameters were set as follows: $\alpha = \{1, ..., 1\}$, $\lambda = 0.001$, $\nu = 0.001$; least-squares estimates for η and C: $\eta = (10.38, -0.24)$, $\rho = 2$, $C = \begin{pmatrix} 0.04 & 0 \\ 0 & 0.24 \end{pmatrix}$, $R = \begin{pmatrix} 0.1 & 10^{-5} \\ 10^{-5} & 0.1 \end{pmatrix}$.

7.4.2 Reductive stepwise method

Pretending that we do not know the correct number of clusters, we start with the $(K+1)$ mixture components, compute parameters that adequately describe the data and then collapse all possible pairs of components. If there is a high probability that the minimum KL distance between the original and all collapsed versions is below a certain threshold (i.e. $c_k = 0.5$ as in Sahu and Cheng (2003)), then $(K+1)$ components describe the data better than the K-component model. If not, then we need to choose a K-component version. After that, we need to test the $(K-1)$-component model in the same fashion.

For each value of K, we let the Gibbs sampler run for 200,000 iterations, discarding the first 100,000 burn-in iterations. The mean distance between the $K = 4$ and $K = 3$ component models was $d^* = 0.14$. Component weights were found to be $0.2963, 0.0185, 0.3888, 0.2964$, and component means and elements of the covariance matrix were estimated as:

$$
\begin{cases}
\mu_1 = \begin{pmatrix} 7.992 \\ 1.003 \end{pmatrix}, \Sigma_1 = \begin{pmatrix} 0.0773 & 2.4 \times 10^{-4} \\ 2.4 \times 10^{-4} & 0.07693 \end{pmatrix}, \\[2mm]
\mu_2 = \begin{pmatrix} -1445 \\ 1127 \end{pmatrix}, \Sigma_2 = \begin{pmatrix} 18410 & -18710 \\ -18710 & 19730 \end{pmatrix}, \\[2mm]
\mu_3 = \begin{pmatrix} 12 \\ -1.005 \end{pmatrix}, \Sigma_3 = \begin{pmatrix} 0.05599 & 8.67 \times 10^{-5} \\ 8.67 \times 10^{-5} & 0.05571 \end{pmatrix}, \\[2mm]
\mu_4 = \begin{pmatrix} 9.99 \\ -0.00359 \end{pmatrix}, \Sigma_4 = \begin{pmatrix} 0.07759 & 3.74 \times 10^{-5} \\ 3.74 \times 10^{-5} & 0.07727 \end{pmatrix},
\end{cases}
\tag{7.21}
$$

These results indicate that $K = 4$ components is probably one too many to describe the Zhou dataset: mean distance between four- and three-component models is 0.14, the second component has less than 2% of observations allocated to it, and values of the mean and variance for the second component are too far from the mean values $\eta = (10.38, -0.24)$ and $C = \begin{pmatrix} 0.04 & 0 \\ 0 & 0.24 \end{pmatrix}$. When the same model was analyzed with $K = 3$, the results were much more promising. The mean distance between the $K = 3$ and $K = 2$ component models was $d^* = 0.14$. Component weights were found to be $0.4, 0.3, 0.31$, and component means and elements of the covariance matrix were estimated as:

$$
\begin{cases}
\mu_1 = \begin{pmatrix} 12 \\ -1.2 \end{pmatrix}, \Sigma_1 = \begin{pmatrix} 0.22 & 1.4 \times 10^{-5} \\ 1.4 \times 10^{-5} & 0.23 \end{pmatrix}, \\[2mm]
\mu_2 = \begin{pmatrix} 10 \\ 0.032 \end{pmatrix}, \Sigma_2 = \begin{pmatrix} 0.13 & 2.3 \times 10^{-5} \\ 2.3 \times 10^{-5} & 0.13 \end{pmatrix}, \\[2mm]
\mu_3 = \begin{pmatrix} 7.9 \\ 0.9 \end{pmatrix}, \Sigma_3 = \begin{pmatrix} 0.12 & 5.2 \times 10^{-6} \\ 5.2 \times 10^{-6} & 0.12 \end{pmatrix}.
\end{cases}
\tag{7.22}
$$

The three-component mixture model correctly estimates means and variances for three components. If we compare the simulation results to the original values, we notice that the simulated values for all variables differ from the original values by less than three values of standard deviation. Absence of label switching in the three- and four-component models may indicate lack of convergence.

Table 7.2   Zhou model. Means and standard deviations of parameters within clusters, as defined by the KLC method.

Cluster	μ_1	SD	μ_2	SD	σ	SDS(σ)	Size
1	7.88	0.46	0.9	0.3587	0.0997	0.000398	15
2	10.04	0.52	0.08	0.3202	0.0998	0.000366	14
3	11.86	0.69	−1.09	0.4738	0.1912	0.204150	21

7.4.3   Zhou model using KLC algorithm

At the first step of the KLC algorithm, we computed subject-specific vectors of parameters (μ_1, μ_2, μ_3), as specified in Section 7.4.1). Estimated values of intercepts and slopes for individual subjects are shown in Table 7.2.

WinBUGS simulation produces similar parameter estimates for the model $\mu_1 = (7.87, 0.93)$, $\sigma_1 = 0.12$, $\mu_2 = (10.03, 0.032)$, $\sigma_2 = 0.13$, $\mu_3 = (11.95, -1.89)$, $\sigma_3 = 0.22$; component weights $\pi = (0.3, 0.3, 0.4)$. KLC misclassifies four subjects (Table 7.3, Figure 7.9), allocating three subjects from the first cluster to the second and one subject from the second cluster to the first.

Taking a closer look at these misclassified elements, we notice that values of the intercept and slope for these elements are between clusters 1–2. Using Pearson's correlation coefficient to cluster the subjects, we do not attain similar accuracy (Table 7.4). Curves $Y_4 \approx 8.244 - 0.0205 \times t$, $Y_5 \approx 8.488 + 0.104 \times t$, $Y_{14} \approx 7.658 + 0.09561 \times t$ have slopes that are smaller than the mean slope of the first cluster (0.9), and hence are classified as members of

Table 7.3   Zhou model. Comparison between "true" cluster membership and KLC-estimated cluster membership.

	True 1	True 2	True 3
Estimated 1	12	1	0
Estimated 2	3	14	0
Estimated 3	0	0	20

Table 7.4   Zhou model. Comparison between "true" cluster membership and the Pearson's correlation coefficient cluster membership.

	True 1	True 2	True 3
Estimated 1	6	4	1
Estimated 2	8	5	9
Estimated 3	1	5	10

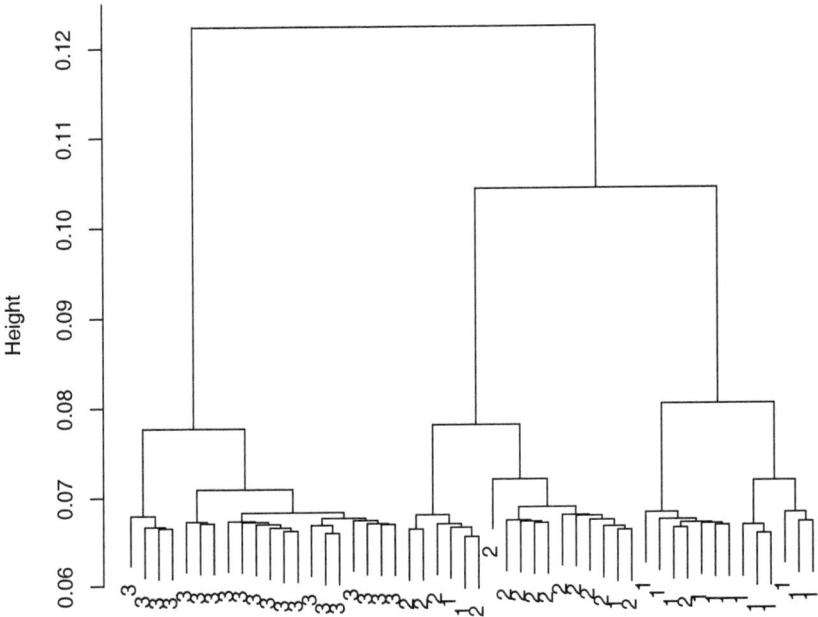

Fig. 7.9. Zhou model. Dendrogram for the Kullback–Leibler clustering, using Kullback–Leibler distance as a similarity measure.

the second cluster; curve $Y_{27} \approx 8.581 + 0.3925 \times t$ has a smaller intercept and a larger slope, which is more characteristic of the first cluster than of the second. As a consequence, these subjects were almost equally likely to be allocated to the first or the second cluster.

7.4.4 Zhou model using KLMCMC

Next we applied KLMCMC (Algorithm 4.2), with parameters $\lambda = \lambda_B = 3$. We ran the simulation for 300 birth–death steps, each containing 5,000 WinBUGS iterations. The resulting posterior distributions of the model parameters are shown in Figures 7.11–7.12.

As we can see from Table 7.5 and Figure 7.12, the solution of the simulated time-series problem using the KLMCMC strategy has advantages over the pure WinBUGS approach, since it produces better estimations of parameters. This can be explained by the lack of WinBUGS

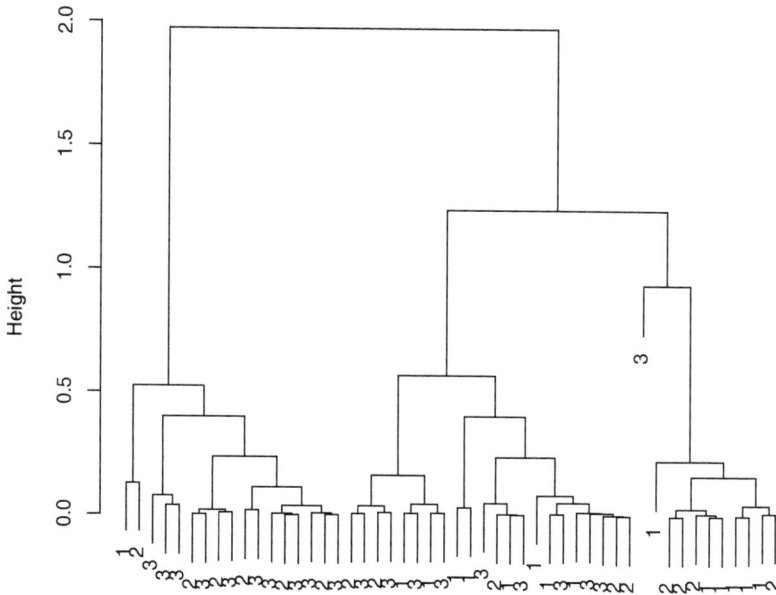

Fig. 7.10.   Zhou model. Dendrogram for the complete linkage clustering, using Pearson's correlation coefficient as a similarity measure.

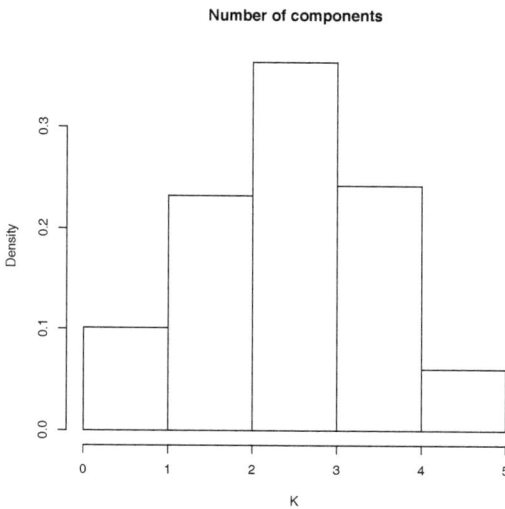

Fig. 7.11.   Zhou model.   Posterior number of components for a simulated time-series problem using KLMCMC with $\lambda = \lambda_B = 3$.

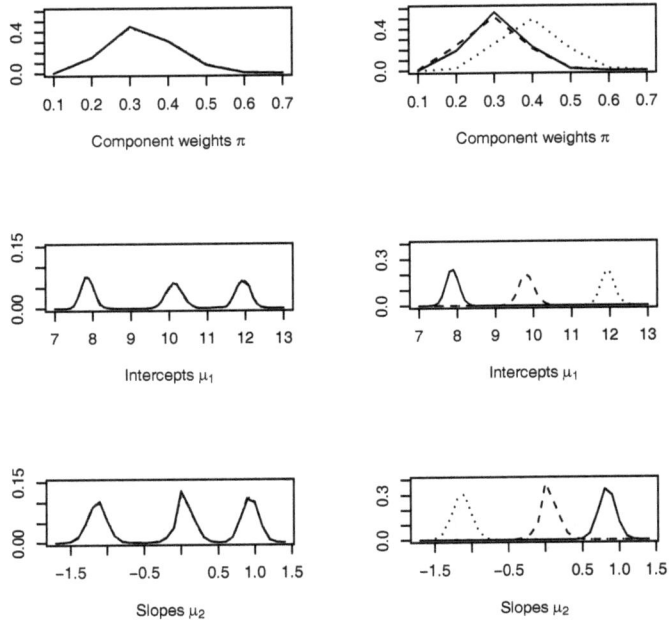

Fig. 7.12.   Zhou model. KLMCMC with random permutation sampler conditional on $K = 3$. Posterior distributions before (left) and after (right) relabeling. Top: component weights, middle: intercepts, bottom: slopes.

Table 7.5   Zhou model. Output of Stephens' relabeling algorithm for the $K = 3$ component model.

Iter	$\hat{w}_1$	$\hat{w}_2$	$\hat{w}_3$	$\hat{\mu}_1^1$	$\hat{\mu}_2^1$	$\hat{\mu}_3^1$
1	0.33	0.33	0.33	10.15	10.15	10.15
2	0.31	0.30	0.39	7.9	10.17	11.93
3	0.31	0.30	0.4	7.9	10.14	11.95
Iter	$\hat{\mu}_1^2$	$\hat{\mu}_2^2$	$\hat{\mu}_3^2$	$\hat{\sigma}_1$	$\hat{\sigma}_2$	$\hat{\sigma}_3$
1	0.04	−0.12	−0.41	0.14	0.14	0.14
2	0.93	0.04	−1.18	0.13	0.13	0.15
3	0.93	0.04	−1.18	0.1	0.11	0.19

convergence due to the stickiness of the Gibbs sampler which is resolved by trans-dimensional methods such as KLMCMC. Additional improvement to convergence is achieved from using the random permutation sampler after each round of WinBUGS in the KLMCMC method.

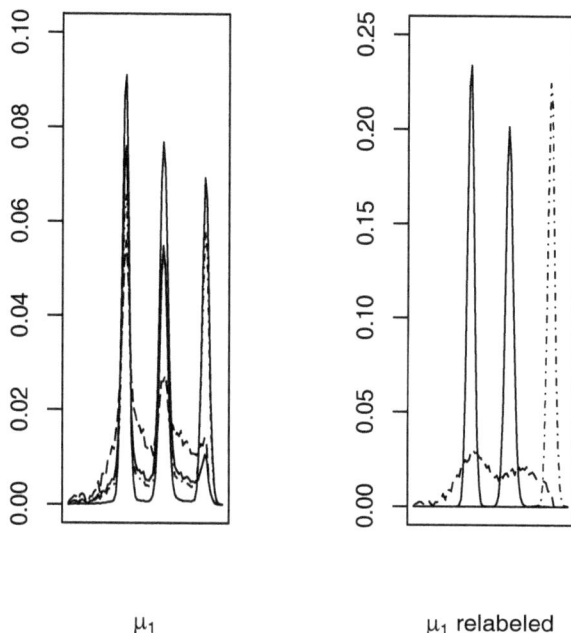

μ_1 μ_1 relabeled

Fig. 7.13. Zhou model. Posterior density of cluster center intercepts for $K = 4$, before and after relabeling according to Stephens' algorithm.

Table 7.6 Zhou model. Output of Stephens' relabeling algorithm for the $K = 4$ component model.

I	$\hat{w}_1$	$\hat{w}_2$	$\hat{w}_3$	$\hat{w}_4$	$\hat{\mu}_1^1$	$\hat{\mu}_2^1$	$\hat{\mu}_1^2$	$\hat{\mu}_2^2$	$\hat{\mu}_1^3$	$\hat{\mu}_2^3$
0	0.23	0.24	0.28	0.25	9.17	11.17	10.43	9.73	0.34	−0.66
1	0.30	0.34	0.27	0.10	7.88	11.98	10.34	10.16	0.93	−1.21
2	0.30	0.33	0.28	0.10	7.87	12.01	10.14	10.70	0.93	−1.23
3	0.30	0.33	0.27	0.10	7.87	12.01	10.11	10.78	0.93	−1.23

I	$\hat{\mu}_1^4$	$\hat{\mu}_2^4$	$\hat{\sigma}_{11}^1$	$\hat{\sigma}_{22}^1$	$\hat{\sigma}_{11}^2$	$\hat{\sigma}_{22}^2$	$\hat{\sigma}_{11}^3$	$\hat{\sigma}_{22}^3$	$\hat{\sigma}_{11}^4$	$\hat{\sigma}_{22}^4$
0	−0.30	−0.02	0.14	0.17	0.16	0.15	0.14	0.17	0.16	0.15
1	−0.10	−0.11	0.13	0.19	0.14	0.17	0.13	0.19	0.14	0.17
2	0.03	−0.48	0.13	0.20	0.14	0.18	0.13	0.20	0.14	0.18
3	0.05	−0.53	0.13	0.20	0.13	0.18	0.13	0.20	0.13	0.18

We also applied Stephens' relabeling algorithm to the output of the Gibbs sampler for w, μ and σ. The posterior density of cluster center intercepts is presented in Figure 7.13 and output of the algorithm is shown in Table 7.6. As can be seen from this plot, the first three components have distinct normal distributions and the fourth component is "smeared". Therefore it is an indication that a smaller number of components is needed for description of this model.

7.5 Transcription Start Sites Prediction

A number of biological terms are used in this section. A legend of terms and abbreviations (adapted from Lodish *et al.* (2000)) follows:

A *eukaryote* is an organism whose cells contain a nucleus and other organelles enclosed within membranes.

Transcription is the first step of gene expression, in which a particular segment of DNA is copied into messenger RNA (mRNA) by RNA polymerase.

A *complementary DNA* (cDNA) is DNA synthesized from a mRNA template. An *expressed sequence tag* (EST) is a sub-sequence of a cDNA sequence.

A *transcription start site* (TSS) is the location where transcription starts at the 5′-end of a gene sequence.

A *promoter* is a regulatory region of DNA usually located upstream of a gene's TSS, providing a control point for regulated gene transcription.

Directionality is the end-to-end chemical orientation of a single strand of nucleic acid. The relative positions of structures along a strand of nucleic acid, including genes and various protein binding sites, are usually noted as being either upstream (towards the 5′-end) or downstream (towards the 3′-end).

The *TATA box* is a DNA sequence found in the promoter region of genes in archaea and eukaryotes.

A *CpG island* is a genomic region that contains a high frequency of CG dinucleotides.

A *transcription factor* is a protein that binds to specific DNA sequences, called *transcription factor binding sites*.

The *start codon* is the first codon of a mRNA translated to form a protein. The start codon always codes for methionine in eukaryotes, and the most common start codon is ATG. Alternate start codons are very rare in eukaryotic genomes.

A *stop codon* is a nucleotide triplet within mRNA that signals a termination of translation. An *open reading frame* (ORF) is the part of a mRNA sequence bounded by start and stop codons.

GenBank is the National Institutes of Health genetic sequence database, an annotated collection of all publicly available DNA sequences.

Accurate determination of TSS of eukaryotic genes is an important task that has not yet been fully solved. Many of the promoter motif-finding methods rely on correct identification of upstream regions and TSS. It was previously reported [Berendzen *et al.* (2006); Pritsker *et al.* (2004); Troukhan *et al.* (2009); Triska *et al.* (2013)] that the position of the transcription factor binding site with respect to the TSS plays a key role in the specific programming of regulatory logic within non-coding regions, highlighting the importance of determining the precise location of the TSS for motif discovery.

A simple approach is to use a collection of 5′ expressed sequence tags (ESTs) to predict the position of the TSS. EST is a short sub-sequence of a transcribed cDNA sequence commonly used for gene discovery [Adams *et al.* (1991); NCBI (2004a)]. One of the problems with the 5′ ESTs is that even in the best libraries only 50–80% of the 5′ ESTs extend to the TSS [Suzuki and Sugano (1997); Sugahara *et al.* (2001); Ohler *et al.* (2002)]. The traditional approach to predicting the position of the TSS was based on finding the position of the longest 5′ transcripts (for overview of eukaryotic promoter prediction methods see Fickett and Hatzigeorgiou (1997)).

Recently, it has been demonstrated that the quality of TSS prediction can be improved when combining data from multiple sources, such as collections of 5′ ESTs and conserved DNA sequence motifs [Ohler *et al.* (2002); Ohler (2006)]. Down and Hubbard (2002) developed a machine-learning approach (Eponine) to build useful models of promoters. Their method uses weight matrices for the most significant motifs around the TSS (e.g. TATA box and CpG island) to predict the positions of the TSS. Eponine was tested on the human chromosome 22 and detected $\sim 50\%$ of experimentally validated TSS, but on some occasions it did not predict the direction of the transcription correctly.

Abeel *et al.* (2009) developed *EP3*, a promoter prediction method based on the structural large-scale features of DNA, including bendability, nucleosome position, free energy, and protein-DNA twist. EP3 identifies the region on a chromosome that is likely to contain the TSS, but it does not predict the direction of a promoter. In 2009 we proposed a different method of TSS prediction that considers positional frequency of 5′ EST matches on genomic DNA together with the gene model, which allows an accurate determination of the TSS. It has been demonstrated that such a statistical approach outperforms deterministic methods [Troukhan *et al.* (2009); Triska *et al.* (2013)]. Additionally, most current methods choose one TSS per locus. However, it has been demonstrated [Joun *et al.* (1997); Tran *et al.* (2002)] that genes can have multiple isoforms and alternative, tissue-specific, TSS.

In order to reliably predict TSS, it is essential to have a good quality assembled genome and a comprehensive collection of ESTs per locus mapped to the potential promoter regions. By 2011 over 70 million ESTs had been accumulated in the NCBI GenBank database. Each EST can be considered as noisy evidence of the positions of TSS. Any genomic position can have from 0 to N (total number of observations) ESTs mapped to it. In the easy case when all N ESTs are mapped to the same position, we have a single reliable prediction of the TSS.

Other cases are more complex. Since each locus may have one or more real TSS, we have a mixture model with an unknown number of components, corresponding to an unknown number of TSS per locus. For illustration, we used the well-annotated genome of *Arabidopsis thaliana*, whose loci have up to 2,500 ESTs per locus mapped to the promoter region. See also Section 5.6. In our application we assumed that the length of the promoter is at most 3,000 nucleotides, and the TSS can be located in any position in the promoter region.

In the framework of our NPB approach, we considered the following model:

$$\theta_i \sim F,$$
$$F \sim DP(\cdot|c, G_0), \ c = 1, \tag{7.23}$$
$$Y_i \sim p(\cdot|\theta_i), \ 1 \leq i \leq N,$$

where G_0 is a uniform distribution on the interval $[0, 1]$. In this case, the equation $Y_i \sim p(\cdot|\theta_i)$, where the density of Y_i is binomial based on n independent trials and where θ_i is the probability of success, N is the number of ESTs corresponding to a given locus. For many genes, there is one major start site, located at a specific position on a genome, and also several minor TSS which may be less precise. Therefore, we have a mixture model setting. Note, that in addition to NPB approach, our approach can utilize nonparametric maximum likelihood framework (called NPEST algorithm), as shown in Tatarinova *et al.* (2014).

Figure 7.14 shows an example of the EST positions, where the major TSS is located around 520 nt upstream from the ATG, and the secondary

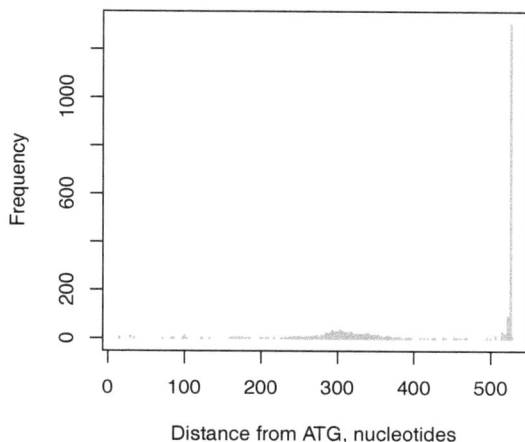

Fig. 7.14. Prediction of TSS: Y, distribution of EST positions.

TSS around 300 nt has larger variability. Both TSS have approximately equal numbers of ESTs attributed to them. The binomial model provides a convenient description of the TSS-prediction problem. We consider each position Y_i on the 1,000 nt long promoter as the number of successes in n Bernoulli trials. Figures 7.15 and 7.16 show the results of a Gibbs sampler, predicting two alternative TSS: at 525 and at 300. This is consistent with our visual examination of the EST distribution in the promoter region.

The model is easy to implement in WinBugs and JAGS. However, the computational time is too long for whole-genome analysis (a locus may take up to 30 minutes of running time). Hence we have developed a parallel version of the method and used high performance computer clusters, since all 30K genes are independent and can be treated in parallel.

A comprehensive collection of TSS prediction for several genomes (such as *Arabidopsis thaliana*, *Homo sapiens* and *Oryza sativa*) is available at the www.glacombio.net. Figure 7.17 shows the result of the prediction of TSS for Arabidopsis locus *At3g26520*. This gene encodes gamma tonoplast intrinsic protein involved in a defense response to bacterium, hydrogen peroxide transmembrane transport, and water homeostasis.

Using our nonparametric approach we predicted TSS for 13,208 loci. We compared our approach to The Arabidopsis Information Resource (TAIR) annotations. To make this comparison we used only the main predictions made by all three methods. We have selected those loci that had at least

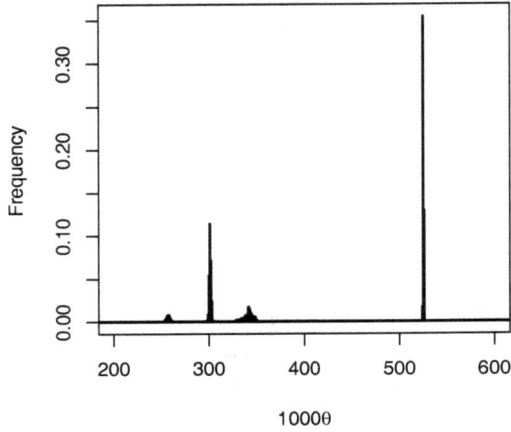

Fig. 7.15.   Prediction of TSS: $F|Y$.

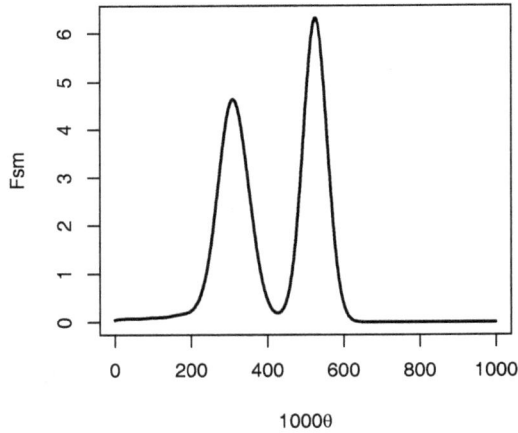

Fig. 7.16.   Prediction of TSS: smoothed $F|Y$.

five EST mapped to the mode of the EST distribution. The number of predicted TSS ranged from 1 to 15, with 63% of analyzed loci having one TSS predicted, 22% had two TSS, 8% had three TSS. Pearson's correlation coefficient between the number of ESTs per locus and the number of predicted TSS is 0.037. Hence there is no significant relationship between the two values.

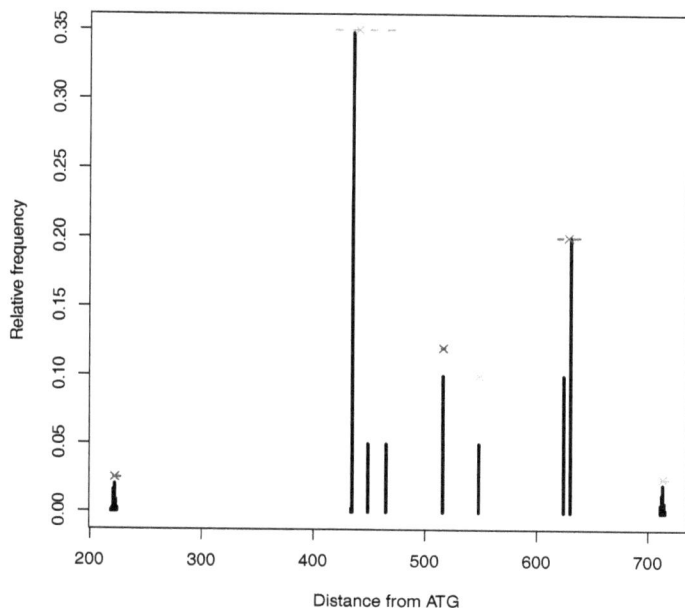

Fig. 7.17. TSS prediction for locus *At3g26520*. Height of vertical lines corresponds to relative frequency of predicted TSS; horizontal lines show 99% confidence intervals for the predicted TSS.

The presence of multiple TSS per locus has a weak association with alternative splicing: those loci that have one open reading frame (ORF) predicted have a 37% chance of multiple TSS, as compared to a 45% chance for those with multiple ORFs. The mode of the distribution of untranslated region lengths was found to be 80 nucleotides.

We use known statistical features of core promoter regions to compare the performance of TSS-prediction methods. The TATA box, located around position −30 from the TSS is the most over-represented sequence pattern. Another conserved feature is the initiator (Inr), located at TSS, commonly containing a dinucleotide sequence of CA. We compared frequencies of a canonical TATA-box four-nucleotide sequence (TATA) in promoters predicted by TAIR and by NPB: 30% of TAIR-predicted promoters contain TATA at positions −40, ..., −20, as compared to 38% of NPB-predicted promoters. Less common four-nucleotide forms of the TATA box (such as TAAA and CTAT) are also more prevalent in NPB-predicted promoters

(61% vs. 55%). In addition, there is stronger nucleotide consensus at TSS (44% of C and 35% of T followed by 63% of A) for NPB than for TAIR (35% of C and 39% of T followed by 53% of A). These observations indicate that NPB has a potential for reliable identification of TSS.

7.6 Conclusions

In this chapter, we presented new methods to analyze gene expression time-series data. We assumed that the underlying biological process responsible for the change of mRNA levels in a cell can be described by a piecewise continuous function. We proposed to approximate the parameters of this function using a multivariate normal distribution. Our methods found parameters of such a function for individual genes and then genes were clustered based on the values of these parameters rather than observed expression. As a result of the clustering procedure, we obtained a smooth centroid curve and a set of curve mean parameters and standard deviations for each cluster. We believe that this approach better reflects biological continuity of cell processes.

Both KLMCMC and KLC are novel methods for time-series observation clustering that can be applied after some initial data filtering and pre-clustering analysis. KLMCMC is a true trans-dimensional method: it determines the optimal number of clusters for a given dataset and selected model. KLC is better suited to handling larger datasets than KLMCMC, but it has a less obvious probabilistic interpretation.

The proposed methods may be used for clustering of time-series data with an unknown number of clusters, providing not only the cluster membership as its output, but also a mathematical model of gene behavior. Although the total number of genes on a genome is computationally prohibitive to be analyzed by WinBUGS, the pre-processing step can reduce the problem to a manageable size by eliminating those genes that do not show differential expression during the experiment of interest.

In addition to cluster membership, KLMCMC and KLC produce a meaningful and easy-to-interpret biological profile. KLMCMC and KLC require careful selection of the model for fitting; they are best suited for refinement of some preliminary clustering. The number of clusters is sensitive to the choice of prior distributions. Hence, biological and experiment-specific insight should be used to determine informative priors. Our methods can be successfully used for cluster refinement and construction of models of cellular processes. Since KLMCMC does not involve likelihood evaluation, it is more computationally efficient as compared to established methods (birth–death Markov chain Monte Carlo and reversible jump Markov chain Monte Carlo) that, in case of nonlinear mixture models, require multiple numerical evaluations of integrals, in case of nonlinear mixture models.

The NPB and NPEST approaches can be efficiently applied to problems outside pharmacokinetics. Our novel algorithm NPB offers prediction of multiple TSS per locus and enhances the understanding of alternative splicing mechanisms. We presented a novel nonparametric method for analysis of EST distributions, NPB. The method was applied to the genome of *Arabidopsis thaliana*, and predicted promoters were compared to TAIR predictions. The proposed method expands recognition capabilities to multiple TSS per locus and enhances the understanding of alternative splicing mechanisms.

Appendix A

Standard Probability Distributions

In this appendix we summarize essentially all the probability distributions that were used in this book. The density and the first two moments are given.

One-dimensional normal distribution: $X \sim N(\cdot|\mu, \sigma^2)$

$$f(x|\mu, \sigma^2) = \frac{1}{\sqrt{2\pi\sigma^2}} e^{-\frac{(x-\mu)^2}{2\sigma^2}}, \qquad (A.1)$$

$$EX = \mu, \, VarX = \sigma^2.$$

Multivariate normal distribution: $X \sim N(\cdot|\mu, \Sigma)$

$$f(x|\mu, \Sigma) = \frac{1}{(2\pi)^{m/2}\sqrt{|\Sigma|}} e^{-\frac{1}{2}(x-\mu)^T \Sigma^{-1}(x-\mu)}, \qquad (A.2)$$

$$EX = \mu, \, CovX = \Sigma,$$

where Σ is an $m \times m$ positive definite matrix.

Gamma distribution: $X \sim G(\cdot|r, \mu)$

$$f(x|r, \mu) = \frac{\mu^r x^{r-1} e^{-\mu x}}{G(r)}, \text{ where } x > 0, \qquad (A.3)$$

$$EX = \frac{r}{\mu}, \, VarX = \frac{r}{\mu^2}.$$

Poisson distribution: $X \sim Pois(\cdot|\lambda)$

$$f(X = k|\lambda) = \lambda^k \frac{exp(-\lambda k)}{k!}, \, k = 0, 1, ..., \qquad (A.4)$$

$$EX = VarX = \lambda.$$

Binomial distribution: $X \sim Bin(\cdot|n, p)$

$$f(X = k|n, p) = C_k^n p^k (1-p)^{n-k}, \, k = 0, ..., n, \qquad (A.5)$$

$$EX = np, \, VarX = np(1-p).$$

Beta distribution: $X \sim Beta(\cdot|\alpha, \beta)$

$$f(x|\alpha, \beta) = \frac{1}{B(\alpha, \beta)} x^{\alpha}(1-x)^{\beta-1}, \, 0 \le x \le 1, \qquad (A.6)$$

$$EX = \frac{\alpha}{\alpha + \beta},$$

$$VarX = \frac{\alpha\beta}{(\alpha + \beta)^2(\alpha + \beta + 1)}$$

Wishart distribution: $V \sim W(\cdot|q, \Psi)$

$$f(V|q, \Psi) = \frac{|V|^{\frac{q-m-1}{2}}}{2^{\frac{qm}{2}}|\Psi|^{\frac{q}{2}}\Gamma_m(\frac{q}{2})} exp\left(-\frac{1}{2}tr(\Psi^{-1}V)\right), \qquad (A.7)$$

$$EV = q\Psi,$$

$$VarV = q(\psi_{ij}^2 + \psi_{ii}\psi_{jj}),$$

where V is $m \times m$ positive definite matrix and $\Gamma_m(\cdot)$ is the multivariate gamma function:

$$\Gamma_m(a) = \pi^{m(m-1)/4} \prod_{j=1}^{m} \Gamma[a + (1-j)/2].$$

Dirichlet distribution: $W \sim Dir(\cdot|\alpha)$

$$f(w|\alpha) = \frac{\Gamma(\sum_{i=1}^{n}\alpha_i)}{\prod_{i=1}^{n}\Gamma(\alpha_i)} \prod_{i=1}^{n} w_i^{\alpha_i-1}, \qquad (A.8)$$

where $w = (w_1, ..., w_n)$, $\alpha = (\alpha_1, ..., \alpha_n)$, $\sum_{i=1}^{n} w_i = 1$, $\alpha_i > 0$ for all i,

$$E[w_i] = \frac{\alpha_i}{\sum_j \alpha_j}, \quad Var[w_i] = \frac{\alpha_i \sum_{j \ne i} \alpha_j}{(\sum_j \alpha_j)^3 + (\sum_j \alpha_j)^2}.$$

Categorical distribution: $X \sim Cat(\cdot|p))$

$$f(X = k|p) = p_k, \qquad (A.9)$$

where $X \in \{1, 2, ..., m)\}, p = (p_1, ..., p_m)$ and $\sum_{i=1}^{m} p_i = 1$.

Uniform distribution: $X \sim U[a, b]$

$$f(x|a, b) = \frac{1}{b-a}, \text{ if } a \le x \le b, \text{ and } 0 \text{ elsewhere.} \qquad (A.10)$$

$$E[X] = \frac{a+b}{2}.$$

$$Var[X] = \frac{(b-a)^2}{12}.$$

Appendix B

Full Conditional Distributions

Consider the equation:

$$Y_i \sim N(\cdot | h_i(\theta_i, \tau^{-1} \Omega_i)$$

with density $\eta(Y_i | h_i(\theta_i), \tau^{-1} \Omega_i)$, where in general $\eta(x|\mu, \Sigma)$ is the normal density with mean vector μ and covariance matrix Σ evaluated at x.

$$\eta(x|\mu, \Sigma) = \frac{1}{(2\pi)^{\frac{m}{2}} \sqrt{det(\Sigma)}} exp\left(-\frac{1}{2}(x-\mu)^T \Sigma^{-1} (x-\mu)^T\right),$$

$$m = dim(x),$$

$$\theta_i \sim N(\cdot | \mu_{Z_i}, \Sigma_{Z_i}),$$

$$Z_i \sim Cat(\cdot | w), w = (w_1, ..., w_K),$$

$$P(Z_i = k | w) = w_k,$$

where

$$(w_1, ..., w_K) \sim Dir(a_1, ..., a_K), \ (Dir = Dirichlet),$$

$$\tau \sim G(a, b), \ (G = Gamma),$$

$$\mu_k \sim N(\cdot | \lambda_k, \Sigma_k), \ (N = Normal),$$

$$(\Sigma_k)^{-1} \sim W(\cdot | q_k, \Psi_k), \ (W = Wishart).$$

To calculate the full conditional distributions, we first need to compute the joint distributions of all the variables:

$$p(Y^N, \theta^N, Z^N, w^K, \mu^K, \Sigma^K, \tau) = p(\tau) \left[\prod_{i=1}^{N} \eta(Y_i | h_i(\theta_i), \tau^{-1} \Omega_i)\right]$$

$$\times \left[\prod_{i=1}^{N} \eta(\theta_i | \mu_{Z_i}, \Sigma_{Z_i})\right] \left[\prod_{i=1}^{N} p(Z_i | w^K)\right] \left[\prod_{k=1}^{K} p(\mu_k) p(\Sigma_k) p(w^k)\right].$$

Now use Bayes' formula $P(u|v) = \frac{P(u,v)}{P(v)} = cP(u,v)$ in Eq. (B.1), where c is a normalizing constant. Then to find the full conditional distribution of any variable, say τ, given all the other variables, written $p(\tau|...)$, just pick out the terms in Eq. (B.1) that involve τ:

$$p(\tau|...) = cG(\tau|a, b) \prod_{i=1}^{N} \eta(Y_i|h_i(\theta_i), \tau^{-1}\Omega_i]$$

$$= c\tau^{a-1}exp(-b\tau)\tau^{\frac{M}{2}}exp\left(-\frac{1}{2}\sum_{i=1}^{N}(Y_i - h_i(\theta_i))^T(\tau^{-1}\Omega_i)^{-1}(Y_i - h_i(\theta))\right).$$

Notice that

$$c\tau^{a-1}exp(-b\tau)\tau^{\frac{M}{2}}exp\left(-\frac{\tau}{2}D\right) = c\tau^{a+\frac{M}{2}-1}exp\left(-\tau(b + \frac{D}{2})\right),$$

$$D = \sum_{i=1}^{N}(Y_i - h_i(\theta_i))^T(\tau^{-1}\Omega_i)^{-1}(Y_i - h_i(\theta)).$$

Hence,

$$p(\tau|...) = cG\left(\tau|a + \frac{M}{2}, b + \frac{D}{2}\right).$$

But $G(\tau|a + \frac{M}{2}, b + \frac{D}{2})$ is already a properly normalized probability density function. Therefore, the normalizing constant $c = 1$ and $p(\tau|...) = G(\tau|a + \frac{M}{2}, b + \frac{D}{2})$. Next,

$$p(\theta_i|...) = c\eta(Y_i|h_i(\theta_i), \tau^{-1}\Omega_i)\eta(\theta_i|\mu_{Z_i}, \Sigma_{Z_i}).$$

The rest follow in the same way:

$$p(w^K|...) = cp(Z^N|w^K)p(w^K) = c\left(\prod_{i=1}^{N}w^{Z_i}\right)w^{a_1-1}...w^{a_K-1},$$

$$p(Z_i = k|...) = cw_k\eta(\theta_i|\mu_k, \Sigma_k),$$

$$p(\mu_k|...) = c\left[\prod_{i=1, Z_i=k}^{N}\eta(\theta_i|\mu_{Z_i}, \Sigma_{Z_i})\right]p(\mu_k|\lambda_k, \Lambda_k),$$

$$p(\Sigma_k^{-1}|...) = c\left[\prod_{i=1, Z_i=k}^{N}\eta(\theta_i|\mu_{Z_i}, \Sigma_{Z_i})\right]p(\Sigma_k^{-1}|q_k, \Psi_k).$$

Further, since the full conditionals of $(w^K, Z_i, \mu_k, \Sigma_k^{-1})$ above are independent of Y^N and τ, these quantities are exactly the same as if θ^N was observable. This is the standard case for Bayesian mixtures and the corresponding full conditionals can be taken from many references, e.g. see Fruhwirth-Schnatter (2006)(page 194). The results are as follows:

$$p(w^K|...) = D(n_1 + a_1, ..., n_K + a_k),$$

$$p(Z_i = k|...) = \frac{w_k \eta(\theta_i|\mu_k, \Sigma_k)}{\sum_{j=1}^{K} w_j \eta(\theta_i|\mu_j, \Sigma_j)},$$

$$p(\mu_k|...) = N(\mu_k|\lambda_k^*, \Lambda_k^*),$$

$$p(\Sigma_k^{-1}|...) = W(\Sigma_k^{-1}|q_k + n_k, (\Psi_k^{-1} + A_k)^{-1}),$$

where

$$n_k = \sum_{i=1}^{N} I(Z_i = k), \text{ number of times } Z_i = k,$$

$$\theta_k^* = \frac{1}{n_k} \sum_{i=1, Z_i=k}^{N} \theta_i,$$

$$\lambda_k = \Lambda_k (n_k \Sigma_k^{-1} \theta_k^* + \Lambda_k^{-1} \lambda_k),$$

$$\Lambda_k^* = (n_k \Sigma_k^{-1} + \Lambda_k^{-1}),$$

$$A_k = \sum_{i=1, Z_i=k} (\theta_i - \theta_k^*)(\theta_i - \theta_k^*)^T.$$

B.1 Binomial/Beta

Consider the binomial distribution with parameters n and p with density $f(x) = C(n, x)p^x(1 - p)^{(n-x)}, x = 0, ..., n; 0 \le p \le 1$. If the sampling distribution for x is $Bin(n, p)$ with n known, and the prior distribution is $Beta(\alpha, \beta)$ with density $f(p) = \frac{\Gamma(\alpha+\beta)}{\Gamma(\alpha)\Gamma(\beta)}p^{\alpha-1}(1 - p)^{\beta-1}$, then the posterior distribution for p can be computed as:

$$f(p|data) \propto \frac{\Gamma(\alpha + \beta)}{\Gamma(\alpha)\Gamma(\beta)}p^{\alpha-1}(1 - p)^{\beta-1} \prod_i^m C(n, x_i)p^{x_i}(1 - p)^{(n-x_i)}$$

$$\propto C \times p^{\alpha - \sum_i^m x_i - 1}(1 - p)^{\beta - 1 - \sum_i^m x_i - mn}.$$

Therefore, the posterior distribution is $Beta(\alpha + \sum_i^m x_i, \beta + m \times n - \sum_i x_i)$.

Appendix C

Computation of the Weighted Kullback–Leibler Distance

C.1 Weighted Kullback–Leibler Distance for a Mixture of Univariate Normals

Consider the mixture of one-dimensional normal densities:

$$F^{(K)}(\cdot) = \sum_{j=1}^{K} w_j f(\cdot|\mu_j, \sigma_j^2),$$

where

$$f(x|\mu, \sigma^2) = \frac{1}{\sqrt{2\pi\sigma^2}} e^{-\frac{(x-\mu)^2}{2\sigma^2}}$$

From the results of Section 4.2, we only have to calculate the distance between the two-component and one-component models. Assume that we have collapsed components $k_1 = 1$ and $k_2 = 2$. Then the weighted Kullback–Leibler distance between the K-component mixture and the collapsed $(K-1)$-component mixture in this case is

$$d^*(F^{(K)}, F^{*(K-1)}_{k_1,k_2}) = d^*(F^{(2)}, F^{*(1)}_{1,2})$$

$$= w_1 d(f(\cdot|\mu_1, \sigma_1^2), f(\cdot|\mu^*, \sigma^{*2}))$$

$$+ w_2 d(f(\cdot|\mu_2, \sigma_2^2), f(\cdot|\mu^*, \sigma^{*2}))$$

where $d(f, g)$ is the Kullback-Leibler distance between the distributions f and g, and where

$$(\mu^*, \sigma^*) = argmin_\phi\{w_1 d(f(\cdot|\phi_1), f(\cdot|\phi)) + w_2 d(f(\cdot|\phi_2), f(\cdot|\phi))\}, \phi = (\mu, \sigma).$$

It is well-known, see Gil *et al.* (2013), that the KL distance between two normal distributions is given by:

$$d(f(\cdot|\mu_1, \sigma_1^2), f(\cdot|\mu_2, \sigma_2^2))$$

$$= (1/2)log\left(\frac{\sigma_2^2}{\sigma_1^2}\right) + \frac{1}{2\sigma_2^2}(\sigma_1^2 + (\mu_2 - \mu_1)^2 - 1).$$

Now make the change of variables $\gamma = 1/\sigma_2^2$ and $\mu = \mu_2$. The last equation then becomes

$$d^*(f(\cdot|\mu_1, \sigma_1^2), f(\cdot|\mu, \gamma))$$
$$= -log(\gamma) + (1/2)\gamma(\sigma_1^2 + (\mu - \mu_1)^2) + C, \qquad (C.1)$$

where C is a term independent of μ and γ. The right hand side of Eq. (C.1) is a convex function of γ and μ and consequently so is the linear combination

$$H(\mu, \gamma) \equiv w_1 d^*(f(\cdot|\mu_1, \sigma_1^2), f(\cdot|\mu, \gamma))$$
$$+ w_2 d^*(f(\cdot|\mu_2, \sigma_2^2), f(\cdot|\mu, \gamma))$$

Therefore the global minimum of $H(\mu, \gamma)$ is found by simply calculating the stationary points of $H(\mu, \gamma)$ with respect to μ and γ.

Thus we need to find values μ^* and γ^* which solve the gradient equations:

$$\frac{\partial H(\mu, \gamma)}{\partial \mu} = 0,$$

$$\frac{\partial H(\mu, \gamma)}{\partial \gamma} = 0.$$

We have

$$\frac{\partial H(\mu, \gamma)}{\partial \mu} = w_1\gamma(\mu - \mu_1) + w_2\gamma(\mu - \mu_2),$$

$$\frac{\partial H(\mu, \gamma)}{\partial \gamma} = -\frac{w_1 + w_2}{\gamma}$$
$$+ \frac{w_1}{2}(\sigma_1^2 + (\mu - \mu_1)^2) + \frac{w_2}{2}(\sigma_2^2 + (\mu - \mu_2)^2).$$

The solutions (μ^*, σ^{*2}) of the gradient equations are then

$$\mu^* = \frac{w_1\mu_1 + w_2\mu_2}{w_1 + w_2},$$

$$\sigma^{*2} = \frac{w_1\sigma_1^2 + w_2\sigma_2^2}{w_1 + w_2} + \frac{w_1 w_2(\mu_1 - \mu_2)^2}{(w_1 + w_2)^2}.$$

(C.2)

Using the relations

$$\mu^* - \mu_2 = w_1 \frac{(\mu_1 - \mu_2)}{(w_1 + w_2)},$$

$$\mu^* - \mu_1 = -w_2 \frac{(\mu_1 - \mu_2)}{(w_1 + w_2)},$$

we have that the minimum distance $d^*(F^{(K)}, F_{k_1,k_2}^{*(K-1)})$ is given by

$$d^*(F_{k=\{1,2\}}^{(2)}, F_{1,2}^{*(1)}) = \sum_{\kappa \in \{1,2\}} \frac{w_\kappa}{2} \{\frac{\sigma_\kappa^2}{\sigma^{*2}} + log(\sigma^*(\sigma_\kappa)^{-1}) - n\}$$

$$+ \frac{w_1 w_2}{2(w_1 + w_2)^2} \frac{(\mu_1 - \mu_2)^2}{\sigma^{*2}}.$$

C.2 Weighted Kullback–Leibler Distance for a Mixture of Multivariate Normals

Consider a mixture of n-dimensional multivariate normal densities:

$$F^{(k)})(\cdot) = \sum_{j=1}^{k} w_j N(\cdot|\mu_j, \Sigma_j),$$

where

$$N(x|\mu, \Sigma) = \frac{1}{(2\pi)^{n/2}\sqrt{\| \Sigma \|}} e^{-\frac{1}{2}(x-\mu)^T \Sigma^{-1}(x-\mu)}$$

and Σ is an $n \times n$ positive definite matrix. From the results of Section 4.2, we only have to calculate the distance between the two-component and one-component models. Assume that we have collapsed components $k_1 = 1$ and $k_2 = 2$. Then the weighted Kullback–Leibler distance between the K-component mixture and the collapsed $(K - 1)$-component mixture in this case is

$$d^*(F^{(K)}, F_{k_1,k_2}^{*(K-1)}) = d^*(F^{(2)}, F_{1,2}^{*(1)})$$

$$= w_1 d(N(\cdot|\mu_1, \Sigma_1), N(\cdot|\mu^*, \Sigma^*)) + w_2 d(N(\cdot|\mu_2, \Sigma_2), N(\cdot|\mu^*, \Sigma^*)),$$

where

$$(\mu^*, \Sigma^*) = argmin_\phi \{w_1 d(N(\cdot|\phi_1), N(\cdot|\phi)) + w_{k_2} d(N(\cdot|\phi_2), N(\cdot|\phi))\},$$

$$\phi = (\mu, \Sigma).$$

It is well-known, see Gil *et al.* (2013), that the KL distance between two multivariate normal distributions is given by:

$$d(N(\cdot|\mu_1, \Sigma_1), N(\cdot|\mu_2, \Sigma_2)) = (1/2)\{trace((\Sigma_2^{-1}\Sigma_1)$$

$$+ (\mu_2 - \mu_1)^T (\Sigma_2)^{-1}(\mu_2 - \mu_1) - log(det\Sigma_1/det\Sigma_2) - n\}. \tag{C.3}$$

Therefore the weighted Kullback–Leibler distance is equal to

$$d^*(F^{(2)}_{k=\{k_1,k_2\}}, F^{*(1)}_{k_1,k_2}) = \sum_{\kappa \in \{1,2\}} \frac{w_\kappa}{2} \{trace((\Sigma^*)^{-1}\Sigma_\kappa)$$

$$+ (\mu^* - \mu_\kappa)^T (\Sigma^*)^{-1}(\mu^* - \mu_\kappa) + log(det(\Sigma^*(\Sigma_\kappa)^{-1})) - n\}.$$

Now make the change of variables $\Gamma = \Sigma_2^{-1}$ and $\mu = \mu_2$. Eq. (C.3) then becomes

$$d(N(\cdot|\mu_1, \Sigma_1), N(\cdot|\mu, \Gamma^{-1})) = (1/2)\{trace((\Gamma\Sigma_1)$$

$$+ (\mu - \mu_1)^T \Gamma(\mu - \mu_1) - log((det\Gamma)(det\Sigma_1)) - n\}. \tag{C.4}$$

The right-hand side of Eq. (C.4) is a convex function of Γ and μ and consequently so is the linear combination

$$H(\mu, \Gamma) \equiv d^*(F^{(2)}_{k=\{k_1,k_2\}}, F^{*(1)}_{k_1,k_2})$$

$$= w_1 d(N(\cdot|\mu_1, \Sigma_1), N(\cdot|\mu, \Gamma))$$

$$+ w_2 d(N(\cdot|\mu_2, \Sigma_2), N(\cdot|\mu, \Gamma))$$

Therefore the global minimum of $H(\mu, \Gamma)$ is found by simply calculating the stationary points of $H(\mu, \Gamma)$ with respect to μ and Γ.

To compute derivatives of $H(\mu, \Gamma)$ we need some calculus rules for vector-matrix differentials, see Petersen and Pedersen (2012):

$$d(x^T Ax) = x^T (A + A^T)dx$$

$$d(AXB) = A(dX)B$$

$$d(trace(XA)) = trace(dXA)$$

$$d(log(det(XB))) = trace(X^{-1}dX).$$

It follows

$$\frac{\partial H(\mu, \Gamma)}{\partial \mu} = \sum_{\kappa \in \{1,2\}} \frac{w_\kappa}{2} \{\Gamma(\mu - \mu_\kappa) + (\mu - \mu_\kappa)^T \Gamma\},$$

$$\frac{\partial H(\mu, \Gamma)}{\partial \Gamma} = \sum_{\kappa \in \{1,2\}} \frac{w_\kappa}{2} \{-\Gamma \Sigma_\kappa \Gamma \tag{C.5}$$

$$+ \Gamma(\mu - \mu_\kappa)(\mu - \mu_\kappa)^T \Gamma + \Gamma\}. \tag{C.6}$$

Setting to zero the partial derivative with respect to μ we obtain

$$0 = \Gamma((w_1 + w_2)\mu - w_1\mu_1 - w_2\mu_2) + ((w_1 + w_2)\mu - w_1\mu_1 - w_2\mu_2)^T \Gamma$$

Thus

$$\mu^* = \frac{w_1\mu_1 + w_2\mu_2}{(w_1 + w_2)}. \tag{C.7}$$

Setting to zero the partial derivative with respect to Γ we obtain

$$0 = \sum_{\kappa \in \{1,2\}} \frac{w_\kappa}{2} \{-\Gamma \Sigma_\kappa \Gamma + \Gamma(\mu - \mu_\kappa)(\mu - \mu_\kappa)^T \Gamma + \Gamma\}.$$

Multiplying both sides of the last equation by Γ^{-1}, we get

$$0 = \sum_{\kappa \in \{1,2\}} \frac{w_\kappa}{2} \{-\Sigma_\kappa + (\mu - \mu_\kappa)(\mu - \mu_\kappa)^T + \Sigma\},$$

where now we have replaced Γ^{-1} by Σ. It follows:

$$(w_1 + w_2)\Sigma = \{w_2\Sigma_2 + w_1\Sigma_1 + w_1(\mu - \mu_1)(\mu - \mu_1)^T$$

$$+ w_2(\mu - \mu_2)(\mu - \mu_2)^T\}$$

Using Eq.(C.5) we have

$$\mu^* - \mu_2 = w_1 \frac{(\mu_1 - \mu_2)}{(w_1 + w_2)},$$

$$\mu^* - \mu_1 = -w_2 \frac{(\mu_1 - \mu_2)}{(w_1 + w_2)}. \tag{C.8}$$

so that

$$(w_1 + w_2)\Sigma = w_2\Sigma_2 + w_1\Sigma_1 + \frac{w_1 w_2^2}{w_1 + w_2}(\mu_1 - \mu_2)(\mu_1 - \mu_2)^T$$

$$+ \frac{w_2 w_1^2}{w_1 + w_2}(\mu_1 - \mu_2)(\mu_1 - \mu_2)^T$$

$$= w_2\Sigma_2 + w_1\Sigma_1 + w_1 w_2(\mu_1 - \mu_2)(\mu_1 - \mu_2)^T.$$

As a result, the expression for Σ^* is

$$\Sigma^* = \frac{w_2\Sigma_2 + w_1\Sigma_1}{(w_1 + w_2)} + \frac{w_1 w_2(\mu_2 - \mu_1)(\mu_2 - \mu_1)^T}{(w_1 + w_2)^2}.$$

Substituting the expressions in Eqs. (C.6) into the formula for the minimum weighted Kullback–Leibler distance we have

$$d^*(F_{k=\{1,2\}}^{(2)}, F_{1,2}^{*(1)}) = \sum_{\kappa \in \{1,2\}} \frac{w_\kappa}{2} \{trace((\Sigma^*)^{-1}\Sigma_\kappa) + log(det(\Sigma^*(\Sigma_\kappa)^{-1})) - n\}$$

$$+ \frac{w_1 w_2}{2(w_1 + w_2)^2}(\mu_1 - \mu_2)^T(\Sigma^*)^{-1}(\mu_1 - \mu_2).$$

C.3   Weighted Kullback–Leibler Distance for a Mixture of Beta Distributions

Another family of distributions that can be treated analytically is the beta distributions. Consider a mixture of beta densities

$$F^{(k)}(\cdot) = \sum_{j=1}^{k} w_j f(\cdot|\alpha_j, \beta_j)$$

where the probability density function of the beta distribution is

$$f(x|\alpha, \beta) = \frac{1}{B(\alpha, \beta)} x^{\alpha-1}(1 - x)^{\beta-1},$$

where $0 \leq x \leq 1$, α and β are positive numbers and $B(\alpha, \beta)$ is the beta function which we write as

$$B(\alpha, \beta) = \Gamma(\alpha)\Gamma(\beta)/\Gamma(\alpha + \beta).$$

From the results of Section 4.2, we only have to calculate the distance between the two-component and one-component models. Assume that we have collapsed components $k_1 = 1$ and $k_2 = 2$. Then the weighted Kullback–Leibler distance between the K-component mixture and the collapsed $(K - 1)$-component mixture in this case is

$$d^*(F^{(K)}, F_{k_1,k_2}^{*(K-1)}) = d^*(F^{(2)}, F_{1,2}^{*(1)})$$

$$= w_1 d(f(\alpha_1, \beta_1), f\alpha, \beta)) + w_2 d(f(\alpha_2, \beta_2), f(\alpha, \beta))$$

where $d(f, g)$ is the Kullback-Leibler distance between the distributions f and g, and where

$$(\alpha, \beta) = argmin_\phi \{w_1 d(f(\phi_1), f(\phi)) + w_2 d(f(\phi_2), f(\phi))\}, \phi = (\alpha, \beta).$$

It is known, see Gil *et al.* (2013), that the KL distance between two beta distributions is given by:

$$d(f(\alpha_1, \beta_1), f(\alpha, \beta) = \Psi(\alpha_1)(\alpha_1 - \alpha) + \Psi(\beta_1)(\beta_1 - \beta)$$

$$+\Psi(\alpha_1 + \beta_1)(\alpha + \beta - (\alpha_1 + \beta_1)) + log\left[\frac{B(\alpha, \beta)}{B(\alpha_1, \beta_1)}\right],$$

where $\Psi(\alpha)$ is the digamma function $\Psi(\alpha) = \Gamma'(\alpha)/\Gamma(\alpha)$.

Now for fixed α_1, α_2, β_1 and β_2, define the function

$$H(\alpha, \beta) = w_1\{\Psi(\alpha_1)(\alpha_1 - \alpha) + \Psi(\beta_1)(\beta_1 - \beta)$$

$$+ \Psi(\alpha_1 + \beta_1)(\alpha + \beta) + logB(\alpha, \beta)\}$$

$$+ w_2\{\Psi(\alpha_2)(\alpha_2 - \alpha) + \Psi(\beta_2)(\beta_2 - \beta)$$

$$+ \Psi(\alpha_2 + \beta_2)(\alpha + \beta) + logB(\alpha, \beta)\} + C$$

where C is a constant independent of α_1, α_2, β_1 and β_2.

It can be shown, see Dragomir *et al.* (2000), that $logB(\alpha, \beta)$ is a convex function of (α, β) and consequently so is $H(\alpha, \beta)$. Therefore the global minimum of $H(\alpha, \beta)$ is found by simply calculating the stationary points of $H(\alpha, \beta)$ with respect to α and β. Thus we need to find values α^* and β^* which solve the gradient equations:

$$\frac{\partial H(\alpha, \beta)}{\partial \alpha} = 0,$$

$$\frac{\partial H(\alpha, \beta)}{\partial \beta} = 0.$$

Now

$$\frac{\partial logB(\alpha, \beta)}{\partial \alpha} = \Psi(\alpha) - \Psi(\alpha + \beta).$$

Therefore

$$\frac{\partial H(\alpha, \beta)}{\partial \alpha} = -w_1\{\Psi(\alpha_1) - \Psi(\alpha_1 + \beta_1)\} - w_2\{\Psi(\alpha_2) - \Psi(\alpha_2 + \beta_2)\}$$

$$+ (w_1 + w_2)\{\Psi(\alpha) - \Psi(\alpha + \beta)\}.$$

The beta function is symmetric with respect to α and β so that

$$\frac{\partial H(\alpha, \beta)}{\partial \beta} = -w_1\{\Psi(\beta_1) - \Psi(\alpha_1 + \beta_1)\} - w_2\{\Psi(\beta_2) - \Psi(\alpha_2 + \beta_2)\}$$
$$+ (w_1 + w_2)\{\Psi(\beta) - \Psi(\alpha + \beta)\}.$$

Therefore to find α^* and β^*, we need to solve numerically the following two equations in two unknowns:

$$\Psi(\beta^*) - \Psi(\alpha^* + \beta^*) = \frac{w_1\{\Psi(\beta_1) - \Psi(\alpha_1 + \beta_1)\} + w_2\{\Psi(\beta_2) - \Psi(\alpha_2 + \beta_2)\}}{w_1 + w_2},$$

$$\Psi(\alpha^*) - \Psi(\alpha^* + \beta^*) = \frac{w_1\{\Psi(\alpha_1) - \Psi(\alpha_1 + \beta_1)\} + w_2\{\Psi(\alpha_2) - \Psi(\alpha_2 + \beta_2)\}}{w_1 + w_2}.$$

Appendix D

BUGS Codes

D.1 BUGS Code for the Eyes Model

BUGS code for the Eyes model in Eq. 2.35:

```
model
{
    for( i in 1 : N ) {
                y[i] ~ dnorm(mu[i], tau)
                mu[i] <- lambda[T[i]]
                T[i] ~ dcat(P[])
        }
        P[1:2] ~ ddirch(alpha[])
        lambda[2]  ~ dnorm(0.0, 1.0E-6)
        lambda[1] ~ dnorm(0.0, 1.0E-6)
        tau ~ dgamma(0.001, 0.001)
    sigma <- 1/sqrt(tau)
}
```

```
# Data
list (y = c(529.0,  530.0,  532.0,  533.1,  533.4,  533.6,
       533.7,  534.1,  534.8,  535.3,  535.4,  535.9,  536.1,
       536.3,  536.4,  536.6,  537.0,  537.4,  537.5,  538.3,
       538.5,  538.6,  539.4,  539.6,  540.4,  540.8,  542.0,
       542.8,  543.0,  543.5,  543.8,  543.9,  545.3,  546.2,
       548.8,  548.7,  548.9,  549.0,  549.4,  549.9,  550.6,
       551.2,  551.4,  551.5,  551.6,  552.8,  552.9,553.2),
    N = 48, alpha = c(1, 1),
    T = c(1, NA, NA, NA, NA, NA, NA, NA, NA,
    NA, NA, NA, NA, NA, NA, NA, NA, NA,
       NA, NA, NA, NA, NA, NA, NA, NA, NA, NA,
       NA, NA, NA, NA, NA, NA, NA, NA, NA, NA,
       NA, NA, NA, NA, NA, NA, NA, NA, 2))
```

```
# Inits
list (lambda = c(540, 540), tau = 0.1)
```

BUGS code for Eq. (3.36)

```
model
{
        for ( i in 1 : N ){
                y [ i ]  ~   dnorm (lambda [ i ], tau )
                lambda [ i ]  <- mu [ Z [ i ] ]
                Z [ i ]  ~  dcat ( P [ ] )
        }
        P [ 1:2 ]  ~  ddirch ( alpha [ ] )
        beta  ~  dnorm ( 0.0, 1.0E-6 ) I ( 0 , )
        mu [ 1 ]  ~  dnorm ( 0.0, 1.0E-6 )
        mu [ 2 ]  <- mu [ 1 ] + beta
        tau  ~  dgamma ( 0.001, 0.001 )
    sigma  <- 1/sqrt ( tau )
}
```

```
#  Initial  conditions
list ( beta=0, mu [ 1 ]=540 )
```

D.2 BUGS Code for the Boys and Girls Example

```
model{
for(i in 1:N)
{
    theta[1,i,1:2]~dmnorm(mean1[], inv_sigma[,])
    theta[2,i,1:2]~dmnorm(mean2[], inv_sigma[,])

    for( j in 1:T)
    {
        D[i,j] ~ dnorm(mu[i,j],tau)
        mu[i,j]<-theta[I[i],i,1]+theta[I[i],i,2]*t[j]
    }
    I[i]~dcat(P[])
}
P[1:2]~ddirch(alpha[])
mean1[1:2]~dmnorm(eta[],C[,])

addon[1:2]~dmnorm(eta[],C[,])
mean2[1]<-mean1[1]+addon[1]
mean2[2]<-mean1[2]+addon[2]
inv_sigma[1:2,1:2]  ~dwish(R[,],2)
tau~dgamma(0.001,0.001)

inv_tau<-1/tau
}
```

Fig. D.1. Two-component BUGS model for the Boys and Girls example.

D.3   BUGS Code for Thumbtack Data

BUGS code for thumbtacks data (binomial/beta model)

```
model{
  for( i in 1 : N ) {
   S[i] ~ dcat(pi[])
   mu[i] <- theta[S[i]]
   x[i] ~ dbin(mu[i],9)
   for (j in 1 : C) {
    SC[i, j] <- equals(j, S[i])
   }
  }
# Constructive DPP
  p[1] <- r[1]
  for (j in 2 : C) {
   p[j]<-r[j]*(1-r[j-1])*p[j -1]/r[j-1]
  }
  p.sum <- sum(p[])
  for (j in 1:C){
# Uniform prior on theta[i]
   theta[j] ~dbeta(1, 1)
   r[j] ~ dbeta(1, alpha)
# Scaling to ensure sum to 1
   pi[j] <- p[j] / p.sum
  }

# Total clusters
  K <- sum(cl[])
  for (j in 1 : C) {
   sumSC[j] <- sum(SC[ , j])
   cl[j] <- step(sumSC[j] -1)
   }
}
```

Thumbtacks data

```
list (x=c(7,  4,  6,  6,  6,  6,  8,  6,  5,  8,  6,  3,  3,  7,  8,
4,  5,  5,  7,  8,  5,  7,  6,  5,  3,  2,  7,  7,  9,  6,  4,  6,  4,
7,  3,  7,  6,  6,  6,  5,  6,  6,  5,  6,  5,  6,  7,  9,  9,  5,  6,
4,  6,  4,  7,  6,  8,  7,  7,  2,  7,  7,  4,  6,  2,  4,  7,  7,  2,
3,  4,  4,  4,  6,  8,  8,  5,  6,  6,  6,  5,  3,  8,  6,  5,  8,  6,
6,  3,  5,  8,  5,  5,  5,  5,  6,  3,  6,  8,  6,  6,  6,  8,  5,  6,
4,  6,  8,  7,  8,  9,  4,  4,  4,  4,  6,  7,  1,  5,  6,  7,  2,  3,
4,  7,  5,  6,  5,  2,  7,  8,  6,  5,  8,  4,  8,  3,  8,  6,  4,  7,
7,  4,  5,  2,  3,  7,  7,  4,  5,  2,  3,  7,  4,  6,  8,  6,  4,  6,
2,  4,  4,  7,  7,  6,  6,  6,  8,  7,  4,  4,  8,  9,  4,  4,  3,  6,
7,  7,  5,  5,  8,  5,  5,  5,  6,  9,  1,  7,  3,  3,  5,  7,  7,  6,
8,  8,  8,  8,  7,  5,  8,  7,  8,  5,  5,  8,  8,  7,  4,  6,  5,  9,
8,  6,  8,  9,  9,  8,  8,  9,  5,  8,  6,  3,  5,  9,  8,  8,  7,  6,
8,  5,  9,  7,  6,  5,  8,  5,  8,  4,  8,  8,  7,  7,  5,  4,  2,  4,
5,  9,  8,  8,  5,  7,  7,  2,  6,  2,  7,  6,  5,  4,  4,  6,  9,  3,
9,  4,  4,  1,  7,  4,  4,  5,  9,  4,  7,  7,  8,  4,  6,  7,  8,  7,
4,  3,  5,  7,  7,  4,  4,  6,  4,  4,  2,  9,  9,  8,  6,  8,  8,  4,
5,  7,  5,  4,  6,  8,  7,  6,  6,  8,  6,  9,  6,  7,  6,  6,  6),
N=320, C=10, alpha=1)
```

D.4 BUGS Code for Eye-Tracking Data

BUGS Code for Eye-Tracking model

```
model{
   for( i in 1 : N ) {
    S[i] ~ dcat(pi[])
    mu[i] <- theta[S[i]]
    x[i] ~ dpois(mu[i])
    for (j in 1 : C) {
     SC[i, j] <- equals(j, S[i])
     }
   }
# Precision Parameter
    alpha <- 1
# Constructive DPP
   p[1] <- r[1]
   for (j in 2 : C) {
    p[j] <- r[j]*(1-r[j-1])*p[j-1]/r[j-1]
   }
   p.sum <- sum(p[])
   for (j in 1:C){
    theta[j] ~ dgamma(A, B)
    r[j] ~ dbeta(1, alpha)
  # Scaling to ensure sum to 1
    pi[j] <- p[j] / p.sum
   }
# Hierarchical prior on theta[i] or preset parameters
    A <- 1; B <- 1
# Total clusters
   K <- sum(cl[])
   for (j in 1 : C) {
    sumSC[j] <- sum(SC[,])
    cl[j] <- step(sumSC[j] -1)
   }
}
```

Data for Eye-Tracking model

```
# Data
list (x=c(0, 0, 0, 0, 0, 0, 0, 0, 0, 0, 0, 0, 0, 0, 0,
    0, 0, 0, 0, 0, 0, 0, 0, 0, 0, 0, 0, 0, 0, 0, 0, 0,
    0, 0, 0, 0, 0, 0, 0, 0, 0, 0, 0, 1, 1, 1, 1, 1,
    1, 1, 1, 1, 1, 1, 1, 1, 1, 2, 2, 2, 2, 2, 2, 2, 2, 2,
    3, 3, 3, 3, 4, 4, 5, 5, 5, 6, 6, 6, 7, 7, 7, 8, 9, 9,
    10, 10, 11, 11, 12, 12, 14, 15, 15, 17, 17, 22, 24,
    34), N=101, C=19)
```

D.5 BUGS Code for PK Example, Nonparametric Bayesian Approach with Stick-Breaking Priors

```
model{
        for( i in 1 : N ) {
                S[i] ~ dcat(pi[])
                mu[i] <- theta[S[i]]
                v[i]<-V[S[i]]
                for(t in 1:T0){
                        x[i,t] ~ dnorm( h[i,t],tau)
        h[i,t]<-(D/v[i])*exp(-mu[i]*time[t])
                }
                for (j in 1 : C) {
                        SC[i, j] <- equals(j, S[i])
                }
        }
tau<-1/(sigma_e*sigma_e)
# Precision Parameter
alpha ~dgamma(0.1, 0.1)
# Constructive DPP
        p[1] <- r[1]
        for (j in 2 : C) {
                p[j]<-r[j]*(1-r[j-1])*p[j -1]/r[j-1]
        }
        p.sum <- sum(p[])
        for (j in 1:C){
# normal prior on theta[i] and V[j]
        theta[j] ~ dnorm(lambda0, 0.1)
        lambda0 ~ dnorm(0.5, 0.1)
        V[j]~dnorm(V0,0.25), V_0~dnorm N(20, 1)
        r[j] ~ dbeta(1, alpha)
        # scaling to ensure sum to 1
        pi[j] <- p[j] / p.sum
        }
        K <- sum(cl[])    # total clusters
        for (j in 1 : C) {
                sumSC[j] <- sum(SC[ , j])
                cl[j] <- step(sumSC[j] -1)}}}
```

Bibliography

Abeel, T., Peer, Y., and Saeys, Y. (2009). Toward a gold standard for promoter prediction evaluation, *Bioinformatics* **25**, pp. 313–320.

Adams, M., Kelley, J., Gocayne, J., Dubnick, M., Polymeropoulos, M., Xiao, H., Merril, C., Wu, A., Olde, B., Moreno, R., Kerlavage, A., McCombie, W., and Venter, J. (1991). Complementary DNA sequencing: Expressed sequence tags and human genome project, *Science* **252**, 5013, pp. 1651–1656.

Aitkin, M. (2001). Likelihood and Bayesian analysis of mixtures, *Statis. Model.* **1**, pp. 287–304.

Akaike, H. (1974). A new look at statistical model identification, *IEEE Trans. Auto. Control* **19**, pp. 716–723.

Aldous, D. (1985). *Exchangeability and related topics*, École d' Éte de Probabilites de Saint–Flour XII, Vol. 1117 (Springer Lecture Notes in Mathematics).

Alexander, D., Novembre, J., and Lange, K. (2009). Fast model-based estimation of ancestry in unrelated individuals, *Genome Res.* **19**, pp. 1655–1664.

Alon, U., Barkai, N., Notterman, D. A., Gish, K., Ybarra, S., Mack, D., and Levine, A. J. (1999). Broad patterns of gene expression revealed by clustering analysis of tumor and normal colon tissues probed by oligonucleotide arrays, *Proc. Natl. Acad. Sci. USA* **96**, pp. 6745–6750.

Alter, O., Brown, P. O., and Botstein, D. (2000). Singular value decomposition for genome–wide expression data processing and modeling, *Proc. Natl. Acad. Sci. USA* **97**, pp. 10101–10106.

Baek, Y. (2006). *An Interior Point Approach to Constrained Nonparametric Mixture Models*, PhD dissertation, University of Washington, thesis supervisor: Professor James Burke.

Bailey, T. (1995). *Discovering Motifs in DNA and Protein Sequences: The Approximate Common Substring Problem*, PhD dissertation, University of California, San Diego.

Bailey, T. and Gribskov, M. (1998). Methods and statistics for combining motif match scores, *J. Comput. Biol.* **5**, pp. 211–221.

Barash, Y. and Friedman, N. (2002). Context-specific Bayesian clustering for gene expression data, *J. Comput. Biol.* **9**, pp. 169–191.

Bayes, T. and Price, R. (1763). An essay towards solving a problem in the doctrine of chances, *Philosophical Transactions of the Royal Society* **53**, pp. 370–418.

Beal, S. and Sheiner, L. (1995). *NONMEM User's Guide*, NONMEM Project Group, University of California, San Francisco.

Beckett, L. and Diaconis, P. (1994). Spectral analysis for discrete longitudinal data, *Adv. Math.* **103**, pp. 107–128.

Bell, B. (2012). Non-parametric population analysis, `moby.ihme.washington.edu/bradbell/`.

Bennett, J. E., Recine-Poon, A., and Wakefield, J. C. (1996). Markov Chain Monte Carlo for nonlinear hierarchical models, in *Markov Chain Monte Carlo in Practice* (Chapman and Hall, London), pp. 339–357.

Berendzen, K. W., Stber, K., Harter, K., and Wanke, D. (2006). Cis-motifs upstream of the transcription and translation initiation sites are effectively revealed by their positional disequilibrium in eukaryote genomes using frequency distribution curves, *BMC Bioinformatics* **7**, 522.

Best, N. G., Tan, K. K., Gilks, W. R., and Spiegelhalter, D. J. (1995). Estimation of population pharmacokinetics using the Gibbs sampler, *J. Pharmacokinet. Biopharm.* **23**, pp. 407–435.

Blackwell, D. and MacQueen, J. (1973). Ferguson distributions via Polya urn schemes, *Ann. Statist.* **1**, 2, pp. 353–355.

Blum, M., Demierre, A., Grant, D. M., Heim, M., and Meyer, U. A. (1991). Molecular mechanism of slow acetylation of drugs and carcinogens in humans, *Proc. Natl. Acad. Sci. USA* **88**, 12, pp. 5237–5241.

Bowmaker, J. K., Jacobs, G. H., Spiegelhalter, D. J., and Mollon, J. D. (1985). Two types of trichromatic squirrel monkey share a pigment in the red–green region, *Vision Res.* **25**, pp. 1937–1946.

Boyd, S. and Vandenberghe, L. (2004). *Convex Optimization* (Cambridge University Press, Cambridge, UK).

Broet, P., Lewin, A., Richardson, S., Dalmasso, C., and Magdelenat, H. (2004). A mixture model-based strategy for selecting sets of genes in multiclass response micorarray experiments, *Bioinformatics* **20**, 16, pp. 2562–2571.

Brooks, S. P. and Gelman, A. (1998). General methods for monitoring convergence of iterative simulations, *J. Comput. Graph. Statist.* **7**, pp. 434–455.

Cappé, O., Robert, C., and Rydén, T. (2003). CT/RJ–Mix: Transdimensional MCMC for Gaussian mixtures (available as C source code), `www.tsi.enst.fr/$\sim$cappe/`.

Carlin, B. (1999). Hierarchical longitudinal modeling, in *Markov Chain Monte Carlo in Practice* (Chapman and Hall, London), pp. 303–315.

Carlin, B. and Chib, S. (1995). Bayesian model choice via Markov Chain Monte Carlo methods, *J. R. Statist. Soc., B* **57**, pp. 473–484.

Casella, G., Robert, C. P., and Wells, M. T. (2004). Mixture models, latent variables and partitioned importance sampling, *Statist. Meth.* **1**, 1–2, pp. 1–18.

Celeux, G., Hurn, M., and Robert, C. P. (2000). Computational and inferential difficulties with mixture posterior distributions, *J. Amer. Statist. Assoc.* **95**, pp. 957–970.

Chen, I. (2006). Nick Patterson: A cold war cryptologist takes a crack at deciphering DNAs deep secrets, `www.nytimes.com/2006/12/12/science/12prof.html`.

Chu, S., DeRisi, J., Eisen, M., Mulholland, J., Botstein, D., Brown, P. O., and Herskowitz, I. (1998). The transcriptional program of sporulation in budding yeast, *Science* **282**, 5389, pp. 699–705.

Chubatiuk, A. (2013). *Nonparametric Estimation of an Unknown Probability Distribution Using Maximum Likelihood and Bayesian Approaches*, PhD dissertation, University of Southern California, Los Angeles, thesis supervisor: Professor Alan Schumitzky.

Clements, M., van Somerena, E., Knijnenburga, T., and Reinders, M. (2007). Integration of known transcription factor binding site information and gene expression data to advance from co-expression to co-regulation, *Genom. Proteom. Bioinform.* **5**, 2, pp. 86–101.

Cofino, A., Cano, R., Sordo, C., and Gutierrez, J. (2002). Bayesian networks for probabilistic weather prediction, in *Proceedings of the 15th European Conference on Artificial Intelligence, ECAI*, pp. 695–700.

Congdon, P. (2001). *Bayesian Statistical Modeling* (Wiley, New York).

Davidian, M. and Giltinan, D. (1995). *Nonlinear Models for Repeated Measurement Data* (Chapman and Hall, New York).

De la Cruz-Mesía, R., Quintana, F., and Marshall, G. (2008). Model-based clustering for longitudinal data, *Comput. Stat. Data Anal.* **52**, 3, pp. 1441–1457.

Devlin, B., Jones, B. L., Bacanu, S.-A., and Roeder, K. (2002). Mixture models for linkage analysis of affected sibling pairs and covariates, *Genetic Epidemiol.* **22**, pp. 52–65.

Diebolt, J. and Robert, C. P. (1990). Bayesian estimation of finite mixture distributions, part ii: Sampling implementation, Tech. rep., Université de Paris VI, L.S.T.A.

Diebolt, J. and Robert, C. P. (1994). Estimation of finite mixture distributions through Bayesian sampling, *J. R. Statist. Soc., B* **56**, pp. 363–375.

Do, K.-A., Muller, P., and Tang, F. (2003). A Bayesian mixture model for differential gene expression, in *Security and Infrastructure Protection: 35th Symposium on the Interface*, Salt Lake City.

Do, K.-A., Muller, P., and Tang, F. (2005). A Bayesian mixture model for differential gene expression, *J. R. Statist. Soc. C (Appl. Statist.)* **54**, 3, pp. 627–644.

Dowe, D. L., Baxter, R. A., Oliver, J. J., and Wallace, C. S. (1998). Point estimation using the Kullback–Leibler loss function and MML, in *Pacific–Asia Conference on Knowledge Discovery and Data Mining (PAKDD98), Lecture Notes in Artificial Intelligence*, Vol. 1394 (Springer-Verlag), pp. 87–95.

Down, T. and Hubbard, T. (2002). Computational detection and location of transcription start sites in mammalian genomic DNA, *Genome Res.* **12**, 3, pp. 458–461.

Dragomir, S. S., Agarwaland, R. P., and Barnett, N. S. (2000). Inequalities for beta and gamma functions via some classical and new integral inequalities, *J. Inequal. Appl.* **5**, pp. 103–165.

Drinkwater, M. J., Parker, Q. A., Proust, D., Sleza, E., and Quintana, H. (2004). The large scale distribution of galaxies in the Shapley supercluster, *Publ. Astron. Soc. Austral.* **21**, pp. 89–96.

Edwards, I. and Aronson, J. (2000). Adverse drug reactions: Definitions, diagnosis, and management, *The Lancet* **356**, 9237, pp. 1255–1259.

Efron, B., Storey, J. D., Tibshirani, R., and Tusher, V. (2001). Empirical Bayes analysis of a microarray experiment, *J. Amer. Statist. Assoc.* **96**, pp. 1151–1160.

Eisen, M. B., Spellman, P. T., Brown, P. O., and Botstein, D. (1998). Cluster analysis and display of genome–wide expression patterns, *Proc. Natl. Acad. Sci. USA* **95**, pp. 14863–14868.

Elhaik, E., Pellegrini, M., and Tatarinova, T. (2014a). Gene expression and nucleotide composition are associated with genic methylation level in *Oryza sativa*, *BMC Bioinformatics* **15**, 23, online.

Elhaik, E., Tatarinova, T., Chebotarev, D., Piras, I., Calo, C., De Montis, A., Atzori, M., Marini, M., Tofanelli, S., Francalacci, P., Pagani, L., Tyler-Smith, C., Xue, Y., Cucca, F., Schurr, T., Gaieski, J., Melendez, C., Vilar, M., Owings, A., Gomez, R., Fujita, R., Santos, F., Comas, D., Balanovsky, O., Balanovska, E., Zalloua, P., Soodyall, H., Pitchappan, R., ArunKumar, G., Hammer, M., Matisoo-Smith, L., and Wells, S. (2014b). Geographic population structure analysis of worldwide human populations infers their biogeographical origins, *Nat. Commun.* **5**, online.

Escobar, M. and West, M. (1995). Bayesian density estimation and inference using mixtures, *J. Amer. Statist. Assoc.* **90**, pp. 557–588.

Escobar, M. and West, M. (1998). Computing nonparametric hierarchical models, in *Practical Nonparametric and Semiparametric Bayesian Statistics* (Springer, New York), pp. 1–16.

Ferguson, T. (1973). A Bayesian analysis of some nonparametric problems, *Ann. Statist.* **1**, pp. 209–230.

Fickett, J. and Hatzigeorgiou, A. (1997). Eukaryotic promoter recognition, *Genome Res.* **7**, 9, pp. 861–878.

Fienberg, S. E. (2006). When did Bayesian inference become "Bayesian"? *Bayesian Anal.* **1**, 1, pp. 1–40.

Fitzgibbon, L. J., Dowe, D. L., and Allison, L. (2002). Message from Monte Carlo, Tech. Rep. 107, School of Computer Science and Software Engineering, Monash University, Clayton, Victoria 3800, Australia.

Fox, B. (1986). Algorithm 647: Implementation and relative efficiency of quasirandom sequence generators, *Transactions on Mathematical Software* **12**, 4, pp. 362–376.

Fraley, C. and Raftery, A. (2002). Model–based clustering, discriminant analysis, and density estimation, *J. Amer. Statist. Assoc.* **97**, pp. 611–631.

Fraley, C. and Raftery, A. (2007). MCLUST: Software for Model–Based Clustering, Density Estimation and Discriminant Analysis, Tech. rep., Department of Statistics, University of Washington, WA.

Fruhwirth-Schnatter, S. (2001). Markov Chain Monte Carlo estimation of classical and dynamic switching and mixture models, *J. Amer. Statist. Assoc.* **96**, 453, pp. 194–209.

Fruhwirth-Schnatter, S. (2006). *Finite Mixture and Markov Switching Models*, Springer Series in Statistics (Springer).

Fuest, C., Niehues, J., and Peichl, A. (2010). The redistributive effects of tax benefit systems in the enlarged EU, *Public Finance Rev.* **38**, pp. 473–500.

Gelman, A. and Rubin, D. B. (1992). Inference from iterative simulations using multiple sequences, *Statist. Sci.* **7**, pp. 457–511.

Geman, S. and Geman, D. (1984). Stochastic relaxation, Gibbs distribution, and Bayesian restoration of images, *IEEE Trans. PAMI-6*, pp. 721–741.

Gershman, S. and Blei, D. (2012). A tutorial on Bayesian nonparametric models, *J. Math. Psychol.* **56**, pp. 1–12.

Ghosh, D. (2004). Mixture models for assessing differential expression in complex tissues using microarray data, *Bioinformatics* **20**, pp. 1663–1669.

Ghosh, D. and Chinnaiyan, A. M. (2002). Mixture modeling of gene expression data from microarray experiments, *Bioinformatics* **18**, 2, pp. 275–286.

Ghosh, J. and Ramamoorthi, R. (2008). *Bayesian Nonparametrics* (Springer, New York).

Gil, M., Alajaji, F., and Linder, T. (2013). Renyi divergence measures for commonly used univariate continuous distributions, *Information Sciences* **249**, pp. 124–131.

Gilks, W. R., Richardson, S., and Spiegelhalter, D. J. (1996). *Markov Chain Monte Carlo in Practice* (Chapman and Hall, London).

Hartigan, J. and Wong, M. (1979). A K–means clustering alrorithm, *Appl. Statist.* **28**, pp. 126–130.

Hastings, W. K. (1970). Monte Carlo sampling methods using Markov Chains and their applications, *Biometrika* **57**, pp. 97–109.

Hornik, K. (2013). The R FAQ, `CRAN.R-project.org/doc/FAQ/R-FAQ.html`.

Huang, Y., Liu, D., and Wu, H. (2006). Hierarchical Bayesian methods for estimation of parameters in a longitudinal HIV dynamic system, *Biometrics* **62**, pp. 413–423.

Ishwaran, H. and James, L. (2001). Gibbs sampling methods for stick-breaking priors, *J. Amer. Statist. Assoc.* **96**, pp. 161–173.

Jeffrey, I., Madden, S., McGettigan, P., Perriere, G., Gulhane, A., and Higgins, D. (2007). Integrating transcription factor binding site with gene expression datasets, *Bioinformatics* **23**, 3, pp. 298–305.

Joun, H., Lanske, B., Karperien, M., Qian, F., Defize, L., and Abou-Samra, A. (1997). Tissue–specific transcription start sites and alternative splicing of the parathyroid hormone (PTH)/PTH–related peptide (PTHrP) receptor gene: A new PTH/PTHrP receptor splice variant that lacks the signal peptide, *Endocrinology* **138**, 4, pp. 1742–1749.

Kalli, M., Griffen, J., and Walker, S. (2011). Slice sampling mixture models, *Statst. Comput.* **21**, pp. 93–105.

Kaufman, L. and Rousseeuw, P. (1990). *Finding Groups in Data: An Introduction to Cluster Analysis* (Wiley-Interscience, New York).

Keles, S., van der Laan, M., Dudoit, S., Xing, B., and Eisen, M. B. (2003).

Supervised detection of regulatory motifs in DNA sequences, *Bioinformatics* **18**, pp. 1167–1175.

Koenker, R. and Mizera, I. (2013). Convex optimization, shape constraints, compound decisions, and empirical Bayes rules, www.econ.uiuc.edu/roger/research/ebayes/ebayes.html.

Kohonen, T. (1990). The self-organizing map, *Proc. IEEE* **78**, 9, pp. 1464–1480.

Kram, B., Xu, W., and Carter, C. (2009). Uncovering the arabidopsis thaliana nectary transcriptome: investigation of differential gene expression in floral nectariferous tissues, *BMC Plant Biol.* **15**, 9:92.

Kullback, S. and Leibler, R. A. (1951). On information and sufficiency, *Annal. Math. Statist.* **22**, 1, pp. 79–86.

Lau, J. W. and Green, P. J. (2007). Bayesian model-based clustering procedures, *J. Comput. Graph. Statist.* **16**, pp. 526–558.

Leary, R., Jelliffe, R., Schumitzky, A., and Guilder, M. V. (2001). An adaptive grid non–parametric approach to population pharmacokinetic/dynamic (PK/PD) population models, in *Proceedings of 14th IEEE symposium on Computer Based Medical Systems*, pp. 389–394.

Li, B. (2006). A new approach to cluster analysis: The clustering-function-based method, *J. R. Statist. Soc. B* **68**, pp. 457–476.

Li, Q., MacCoss, M., and Stephens, M. (2010). A nested mixture model for protein identification using mass spectrometry, *Annals Appl. Statist.* **4**, 2, pp. 962–987.

Lindsay, B. (1983). The geometry of mixture likelihoods: A general theory, *Ann. Statist.* **11**, pp. 86–94.

Liu, J. S. (1996). Nonparametric hierarchical Bayes via sequential imputations, *Annals Statis.* **24**, 3, pp. 911–930.

Lodish, H., Berk, A., Zipursky, S., Matsudaria, P., Baltimore, D., and Darnell, J. (2000). *Molecular Cell Biology* (W.T. Freeman, New York).

Lopes, H. (2000). *Bayesian Analysis in Latent Factor and Longitudinal Models*, PhD dissertation, Duke University.

Lopes, H., Mueller, P., and Rosner, G. (2003). Meta-analysis for longitudinal data models using multivariate mixture priors, *Biometrics* **66**, 1, pp. 66–75.

Loreda, T. (1995). *From Laplace to Supernova SN 1987A: Bayesian Inference in Astrophysics*, PhD dissertation, University of Chicago.

Lu, X. and Huang, Y. (2014). Bayesian analysis of nonlinear mixed-effects mixture models for longitudinal data with heterogeneity and skewness, *Statist. Med.* **33**, pp. 2830–2849.

Lunn, D., Best, N., Thomas, A., Wakefield, J., and Spiegelhalter, D. (2002). Bayesian analysis of population PK/PD models: general concepts and software, *J. Pharmacokinet. Pharmacodynam.* **29**, pp. 271–307.

Lunn, D., Spiegelhalter, D., Thomas, A., and Best, N. (2009). The BUGS project: Evolution, critique and future directions, *Statist. Med.* **28**, 25, pp. 3049–3067.

MacEachern, S. and Mueller, P. (1998). Estimating mixture of Dirichlet process models, *J. Comput. Graph. Statist.* **7**, 2, pp. 223–238.

MacQueen, J. (1967). Some methods for classification and analysis of multivariate observations, in *Proceedings of the Fifth Berkeley Symposium on Mathematics, Statistics and Probability*, Vol. 1, pp. 281–297.

Mallet, A. (1986). A maximum likelihood estimation method for random coefficient regression models, *Biometrika* **73**, pp. 645–656.

Marin, J.-M., Mengersen, K., and Robert, C. (2005). Bayesian modeling and inference difficulties on mixtures of distributions, in *Handbook of Statistics 25* (Elsevier-Sciences).

Martella, F. (2006). Classification of microarray data with factor mixture models, *Bioinformatics* **22**, 2, pp. 202–208.

McLachlan, G. (1999). Emmix, www.maths.uq.edu.au/$\sim$gjm/.

McLachlan, G., Bean, R., and Peel, D. (2002). A mixture model-based approach to the clustering of microarray expression data, *Bioinformatics* **18**, pp. 413–422.

McLachlan, G., Peel, D., and Bean, R. (2001). Mixture model-based clustering of microarray expression data, in *Australasian Biometrics and New Zealand Statistical Association Joint Conference 2001*.

McLachlan, G. J. and Peel, D. (2000). *Finite Mixture Models*, Wiley Series in Probability and Statistics (John Wiley and Sons, New York).

McLachlan, G. J., Peel, D., and Whiten, W. J. (1996). Maximum likelihood clustering via normal mixture models, *Signal Proc. Image Commun.* **8**, 2, pp. 105–111.

Medvedovic, M. and Sivaganesan, S. (2002). Bayesian infinite mixture model based clustering of gene expression profiles, *Bioinformatics* **18**, pp. 1194–1206.

Medvedovic, M., Yeung, K. Y., and Bumgarner, R. E. (2004). Bayesian mixture model based clustering of replicated microarray data, *Bioinformatics* **20**, pp. 1222–1232.

Mengersen, K. L., Robert, C. P., and Titterington, M. (2011). *Mixtures: Estimation and Applications* (Wiley, Chichester).

Metropolis, N., Rosenbluth, A., Rosenbluth, M., Teller, A., and Teller, E. (1953). Equation of state calculations by fast computing machines, *J. Chem. Phys.* **21**, pp. 1087–1092.

Moen, P. (2000). *Attribute, Event Sequence, and Event Type Similarity Notions for Data Mining*, PhD dissertation, University of Helsinki.

Mueller, P. (1990). A Generic Approach to Posterior Integration and Gibbs Sampling, Technical report no. 91–90, Department of Statistics, Purdue University.

Mueller, P. and Rosner, G. (1997). A semiparametric Bayesian population model with hierarchical mixture priors, *J. Amer. Statist. Assoc.* **92**, pp. 1279–1292.

NCBI (2004a). ESTs: Gene discovery made easier, www.ncbi.nlm.nih.gov/About/primer/est.html.

NCBI (2004b). One size does not fit all: The promise of pharmacogenomics, www.ncbi.nlm.nih.gov/About/primer/pharm.html.

Neal, R. (1996). Sampling from multimodal distributions using tempered transitions, *Statist. Comput.* **6**, 4, pp. 353–366.

Neely, M., van Guilder, M., Yamada, W., Schumitzky, A., and Jelliffe, R. (2012). Accurate detection of outliers and subpopulations with Pmetrics, a nonparametric and parametric pharmacometric modeling and simulation package for R, *Ther. Drug Monit.* **34**, 4, pp. 467–476.

Neyman, J. (1937). Outline of a theory of statistical estimation based on the classical theory of probability, *Phil. Trans. R. Soc. A* **236**, pp. 333–380.

Oh, S.-J. and Kim, J.-Y. (2004). A hierarchical clustering algorithm for categorical sequence data, *Inf. Proc. Lett.* **91**, pp. 135–140.

Ohler, U. (2006). Identification of core promoter modules in Drosophila and their application in accurate transcription start site prediction, *Nucleic Acids Res.* **34**, 20, pp. 5943–5950.

Ohler, U., Liao, G., Niemann, H., and Rubin, G. (2002). Computational analysis of core promoters in the Drosophila genome, *Genome Biol.* **3**, 12.

Pan, W. (2002). A comparative review of statistical methods for discovering differentially expressed genes in replicated microarray experiments, *Bioinformatics* **18**, 4, pp. 546–554.

Papaspiliopoulos, O. and Roberts, G. (2008). Retrospective Markov Chain Monte Carlo methods for Dirichlet process hierarchical models, *Biometrika* **95**, pp. 169–186.

Papastamoulis, P. and Iliopoulos, G. (2009). Reversible jump MCMC in mixtures of normal distributions with the same component means, *Comput. Statist. Data Anal.* **53**, 4, pp. 900–911.

Pauler, D. K. and Laird, N. (2000). A mixture model for longitudinal data with application to assessment of noncompliance, *Biometrics* **56**, pp. 464–472.

Petersen, K. B. and Pedersen, M. S. (2012). The matrix cookbook, http://matrixcookbook.com.

Phillips, D. and Smith, A. (1996). Bayesian model comparison via jump diffusions, in *Markov Chain Monte Carlo in Practice* (Chapman and Hall, London).

Pilla, R. S., Bartolucci, F., and Lindsay, B. G. (2006). Model building for semiparametric mixtures, *The Journal of Statistical Computation and Simulation* **59**, pp. 295–314.

Pirmohamed, M., James, S., Meakin, S., Green, C., Scott, A., Walley, T., Farrar, K., Park, B., and Breckenridge, A. (2004). Adverse drug reactions as cause of admission to hospital: prospective analysis of 18 820 patients, *BMJ* **329**, 7456, pp. 15–9.

Pitman, J. (1996). Some developments of the Blackwell–MacQueen urn scheme, in *Statistics, Probability and Game Theory*, Vol. 30 (IMS Lecture Notes – Monograph Series), pp. 245–267.

Plummer, M. (2003). JAGS: A program for analysis of Bayesian graphical models using Gibbs sampling, in *Proceedings of the 3rd International Workshop on Distributed Statistical Computing (DSC 2003)* (Vienna, Austria).

Pólya, G. (1957). *How to Solve It: A New Aspect of Mathematical Method*, 2nd edn. (Princeton University Press, Princeton).

Postman, M., Huchra, J. P., and Geller, M. J. (1986). Probes of large scale structure in the Corona Borealis region, *Astron. J.* **92**, pp. 1238–1247.

Potthoff, R. F. and Roy, S. N. (1964). A generalized multivariate analysis of variance model useful especially for growth curve problems, *Biometrika* **51**, pp. 313–326.

Pritchard, J. K., Stephens, M., and Donnelly, P. (2000). Inference of population structure using multilocus genotype data, *Genetics* **155**, pp. 945–959.

Pritsker, M., Liu, Y.-C., and Beer, M. (2004). Whole-genome discovery of transcription factor binding sites by network-level conservation, *Genome Res.* **14**, pp. 99–108.

Raftery, A. and Lewis, S. (1992). One long run with diagnostics: Implementation strategies for Markov Chain Monte Carlo, *Statist. Sci.* **7**, pp. 493–497.

Rahnenfuhrer, J. and Futschik, A. (2003). Cost-effective screening for differentially expressed genes in microarray experiments based on normal mixtures, *Austrian J. Statist.* **32**, 3, pp. 225–238.

Ranganathan, A. (2006). The Dirichlet process mixture (DPM) model, `citeseerx.ist.psu.edu/viewdoc/download?doi=10.1.1.133.6665&rep=rep1&type=pdf`.

Raychaudhuri, S., Stuart, J. M., and Altman, R. B. (2000). Principal components analysis to summarize microarray experiments: Application to sporulation time series, in *Pacific Symposium on Biocomputingc* (Hawaii), pp. 452–463.

Richardson, S. and Green, P. J. (1997). On Bayesian analysis of mixtures with an unknown number of components, *J. R. Statist. Soc.* **62**, 1, pp. 731–792.

Riley, S. and Ludden, T. (2005). Pharmacokinetic model selection within a population model using NONMEM and WinBUGS, in *Advanced Methods of PK/PD Systems Analysis* (BMSR/USC Workshop, Los Angeles, CA, organized by D.Z. D'Argenio).

Robert, C. P. (1996). Mixtures of distributions: Inference and estimation, in *Markov Chain Monte Carlo in Practice* (Chapman and Hall, London), pp. 441–464.

Robert, C. P. and Casella, G. (2004). *Monte Carlo Statistical Methods*, 2nd edn. (Springer, New York).

Robert, C. P. and Casella, G. (2011). A short history of Markov chain Monte Carlo: Subjective recollections from incomplete data, *Statist. Sci.* **26**, pp. 102–115.

Roeder, K. (1990). Density estimation with confidence sets exemplified by superclusters and voids in the galaxies, *J. Astron. Soc. Amer.* **85**, 617–624.

Roeder, K. and Wasserman, L. (1997). Practical Bayesian density estimation using mixtures of normals, *J. Amer. Statist. Assoc.* **92**, pp. 894–902.

Rosner, G. and Mueller, P. (1997). Bayesian population pharmacokinetics and pharmacodynamic analyses using mixture models, *J. Pharmacokinet. Biopharm.* **25**, 2, pp. 209–233.

Sahu, S. K. and Cheng, R. (2003). A fast distance based approach for determining the number of components in mixtures, *Canadian J. Statist.* **31**, pp. 3–22.

Schumitzky, A. (1991). Nonparametric EM algorithms for estimating prior distributions, *Appl. Math. Comput.* **45**, pp. 143–157.

Schwarz, G. (1978). Estimating the dimension of a model, *Ann. Statist.* **6**, pp. 730–418.

Sethuraman, J. (1994). A constructive definition of Dirichlet priors, *Statistica Sinica* **4**, pp. 639–650.

Siegmund, K., Laird, P. W., and Laird-Offringa, I. A. (2004). A comparison of cluster analysis methods using DNA methylation data, *Bioinformatics* **20**, 12, pp. 1896–1904.

Smith, B. J. (2005). *Bayesian Output Analysis Program (BOA) User's Manual*, Department of Biostatistics, The University of Iowa College of Public Health.

Spiegelhalter, D., Best, N., Lunn, D., and Thomas, A. (2011). *Bayesian Analysis using BUGS: A Practical Introduction* (Chapman and Hall/CRC, London).

Spiegelhalter, D., Thomas, A., Best, N., and Lunn, D. (2003). *WinBUGS User Manual*, MRC Biostatistics Unit, Institute of Public Health, version 1.4 edn.

Spiegelhalter, D., Thomas, A., Best, N., and Lunn, D. (2006). WinBUGS, `www.mrc-bsu.cam.ac.uk/bugs`, version 1.4.

Spiegelhalter, D. J., Best, N. G., Carlin, B. P., and van der Linde, A. (2002). Bayesian measures of model complexity and fit (with discussion), *J. R. Statist. Soc. B* **34**, 4, pp. 583–639.

Stephens, M. (1997a). *Bayesian Methods for Mixtures of Normal Distributions*, PhD dissertation, University of Oxford.

Stephens, M. (1997b). Contribution to the discussion of paper by Richardson and Green, *J. R. Statist. Soc. B* **59**, pp. 768–769.

Stephens, M. (2000a). Bayesian analysis of mixture models with an unknown number of components – an alternative to reversible jump methods, *Ann. Statist.* **28**, pp. 40–74.

Stephens, M. (2000b). Dealing with label switching in mixture models, *J. R. Statist. Soc. B* **62**, pp. 795–809.

Stephens, M. and Balding, D. J. (2009). Bayesian statistical methods for genetic association studies, *Nature Reviews Genetics* **10**, pp. 681–690.

Sugahara, Y., Carninci, P., Itoh, M., *et al.* (2001). Comparative evaluation of 5'-end-sequence quality of clones in CAP trapper and other full-length-cDNA libraries, *Gene* **263**, pp. 1–2.

Suzuki, Y. and Sugano, S. (1997). Generation of the 5' EST using 5'-end enriched cDNA library, *Tanpakushitsu Kakusan Koso* **42**, 17, pp. 2836–2843.

Tamayo, P., Slonim, D., Mesirov, J., Zhu, Q., Kitareewan, S., Dmitrovsky, E., Lander, E., and Golub, T. (1999). Interpreting patterns of gene expression with self–organizing maps: Methods and application to hematopoietic differentiation, *Proc. Natl. Acad. Sci. USA* **96**, pp. 2907–2912.

Tatarinova, T. (2006). *Bayesian Analysis of Linear and Nonlinear Mixture Models*, PhD dissertation, University of Southern California, thesis supervisor: Professor Alan Schumitzky.

Tatarinova, T., Bouck, J., and Schumitzky, A. (2008). Kullback–Leibler Markov Chain Monte Carlo: A new algorithm for finite mixture analysis and its application to gene expression data, *J. Bioinform. Comput. Biol.* **6**, pp. 727–746.

Tatarinova, T., Kryschenko, A., Triska, M., Hassan, M., Murphy, D., Neely,

M., and Schumitzky, A. (2014). NPEST: a nonparametric method and a database for transcription start site prediction, *Quantitative Biology* **1**, 4, pp. 261–271.

Tatarinova, T., Neely, M., Bartroff, J., van Guilder, M., Yamada, W., Bayard, D., Leary, R., Chubatiuk, A., Jelliffe, R., and Schumitzky, A. (2013). Two general methods for population pharmacokinetic modeling: Non–parametric adaptive grid and non–parametric Bayesian, *J. Pharamacokinet. Pharamacodynam.* **40**, pp. 189–198.

Teh, Y. W. (2010). Dirichlet process, in *Encyclopedia of Machine Learning* (Springer, New York).

Thalamuthu, A., Mukhopadhyay, I., Zheng, X., and Tseng, G. C. (2006). Evaluation and comparison of gene clustering methods in microarray analysis, *Bioinformatics* **22**, pp. 2405–2412.

Tran, P., Leclerc, D., Chan, M., Pai, A., Hiou-Tim, F., Wu, Q., Goyette, P., Artigas, C., Milos, R., and Rozen, R. (2002). Multiple transcription start sites and alternative splicing in the methylenetetrahydrofolate reductase gene result in two enzyme isoforms, *Mamm. Genome* **13**, 9, pp. 483–492.

Triska, M., Grocutt, D., Southern, J., Murphy, D., and Tatarinova, T. (2013). cisExpress: Motif detection in DNA sequences, *Bioinformatics* **29**, 17, pp. 2203–2205.

Troukhan, M., Tatarinova, T., Bouck, J., Flavell, R., and Alexandrov, N. (2009). Genome-wide discovery of cis-elements in promoter sequences using gene expression data, *OMICS: A Journal of Integrative Biology* **13**, pp. 139–151.

Tseng, G. and Wong, W. (2005). Tight clustering: A resampling-based approach for identifying stable and tight patterns in data, *Biometrics* **61**, pp. 10–16.

Wakefield, J. and Walker, S. (1997). Bayesian nonparametric population models: Formulation and comparison with likelihood approaches, *J. Pharmacokinet. Biopharm.* **25**, pp. 235–253.

Wakefield, J. and Walker, S. (1998). Population models with a nonparametric random coefficient distribution, *Sankhya B* **60**, pp. 196–212.

Wakefield, J. C., Smith, A. F. M., Racine-Poon, A., and Gelfand, A. (1994). Bayesian analysis of linear and non-linear population models using the Gibbs sampler, *Appl. Statist.* **34**, 1, pp. 201–221.

Wakefield, J. C., Zhou, C., and Self, S. (2003). Modeling gene expression over time: Curve clustering with informative prior distributions, in *Bayesian Statistics 7* (Oxford University Press, Oxford), pp. 721–732.

Walker, S. (2007). Sampling the Dirichlet mixture model with slices, *Commn. Statist. Simul. Comput.* **36**, pp. 45–54.

Wang, A., Schumitzky, A., and D'Argenio, D. (2007). Nonlinear random effects mixture models: Maximum likelihood estimation via the *EM* algorithm, *Comput. Stat. Data Anal.* **51**, pp. 6614–6623.

Wang, A., Schumitzky, A., and D'Argenio, D. (2009). Population pharmacokinetic/pharmacodyanamic mixture models via maximum *a posteriori* estimation, *Comput. Stat. Data Anal.* **53**, pp. 3907–3915.

Wang, L. and Dunson, D. (2011). Bayesian isotonic density regression, *Biometrika* **98**, pp. 537–551.

Wang, Y. (2007). On fast computation of the non-parametric maximum likelihood estimate of a mixing distribution, *J. R. Statist. Soc. B 69, Part 2, pp. 185-198* **69**, pp. 185–198.

Weigel, D., Altmann, T., and Nover, L. (2004). Atgenexpress, `web.uni0frankfurt.de/fb15/botanik/mcb/AFGN/atgenex.htm`.

Xing, G. and Pilla, R. S. (2004). Flexible fitting and diagnostics of finite mixture models, in *Joint Statistical Meetings: Statistics as a Unified Discipline* (Toronto, Canada).

Yeung, K. Y., Fraley, C., Murua, A., Raftery, A. E., and Ruzzo, W. L. (2001). Model-based clustering and data transformations for gene expression data, *Bioinformatics* **17**, pp. 977–987.

Yeung, K. Y. and Ruzzo, W. L. (2000). An Empirical Study on Principal Component Analysis for Clustering Gene Expression Data, Tech. Rep. UW–CSE–00–11–03, Department of Computer Science and Engineering, University of Washington.

Zhou, C. (2004). *A Bayesian Model for Curve Clustering with Applications to Gene Expression Data Analysis*, PhD dissertation, University of Washington, thesis supervisor: Professor Jon Wakefield.

Zhou, C. and Wakefield, J. (2005). A Bayesian mixture model for partitioning gene expression data, *Biometrics* **62**, 2, pp. 515–525.

Index